ENERGY SCIENCE, ENGINEERING AND TECHNOLOGY

RENEWABLE ENERGY IN CHINA

ENERGY SCIENCE, ENGINEERING AND TECHNOLOGY

Additional books in this series can be found on Nova's website under the Series tab.

Additional e-books in this series can be found on Nova's website under the e-book tab.

ENERGY SCIENCE, ENGINEERING AND TECHNOLOGY

RENEWABLE ENERGY IN CHINA

JIAHAI YUAN
GUOLIANG LUO
AND
JIANGUO CHEN
EDITORS

nova publishers
New York

For permission to use material from this book please contact us:
Telephone 631-231-7269; Fax 631-231-8175
Web Site: http://www.novapublishers.com

Library of Congress Cataloging-in-Publication Data

ISBN: 978-1-63117-610-4

Published by Nova Science Publishers, Inc. † New York

CONTENTS

TABLE LIST

FIGURE LIST

ACKNOWLEDGMENTS

This book is partly supported by Beijing Higher Education Young Elite Teacher Project and the Fundamental Research Funds for the Central Universities. The authors are also grateful to Nadya Gotsiridze-Columbus, whose invitation lead us to discuss and work out the idea of publishing a book on China's renewable energy, to Carra Feagaiga, whose kind reminders helped to finish the preparation of the manuscript in due time, and to the editorial team of Nova, whose editorial services largely improved the manuscript. The errors which remain in the book belong to the authors, certainly.

In: Renewable Energy in China
Editors: J. Yuan, G. Luo and J. Chen

ISBN: 978-1-63117-610-4

Chapter 1

RENEWABLE ENERGY IN CHINA: INTRODUCTION

***Jiahai Yuan*[1*] *and Huiming Xu*[2]**
[1]School of Economics and Management, North China Electric Power University
[2]Information and Telecommunication Branch,
State Grid Corporation of China, People's Republic of China

ABSTRACT

Renewable energy (RE) has grown rapidly in China ever since the publication of the *Renewable Energy Law* in 2006. As the introduction of the book, this chapter begins with an overview of economic growth, energy consumption and CO_2 emissions in China, with a purpose to justify the urgent requirement to develop RE in China's overall picture of social-economic growth. Then the resource endowment of various forms of renewable energy in China is briefly analyzed, and China's legislation and overall plan to develop RE is presented. Finally, with an institutional perspective, the social-political factors of energy system (with an emphasis on power system) development in China are briefly addressed. The chapter concludes with the scope and layout of the entire book.

Keywords: Renewable energy resource, legislation, planning, institutional analysis, China

1. ECONOMIC GROWTH, ENERGY CONSUMPTION AND CARBON EMISSIONS IN CHINA

1.1. Historical Trend

Since implementing the reform-and-open-up policy in1978, China has successfully realized the economic take-off and sustained a growth miracle of successive GDP growth at nearly 10% for more than three decades (Figure 1-1). In 2010, China became the second largest economy with a total GDP of 5878.6 billion US$ (in 2012-year exchange rate and hereafter in this chapter) [1].

* Corresponding author: Email: yuanjh126@126.com.

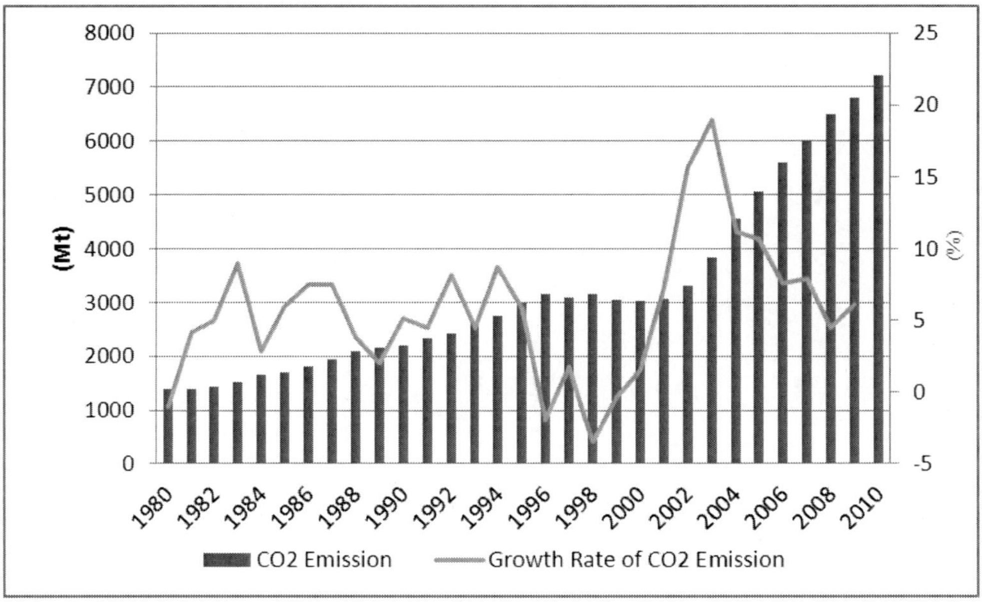

Figure 1-1. GDP and its growth in China, 1980-2011.

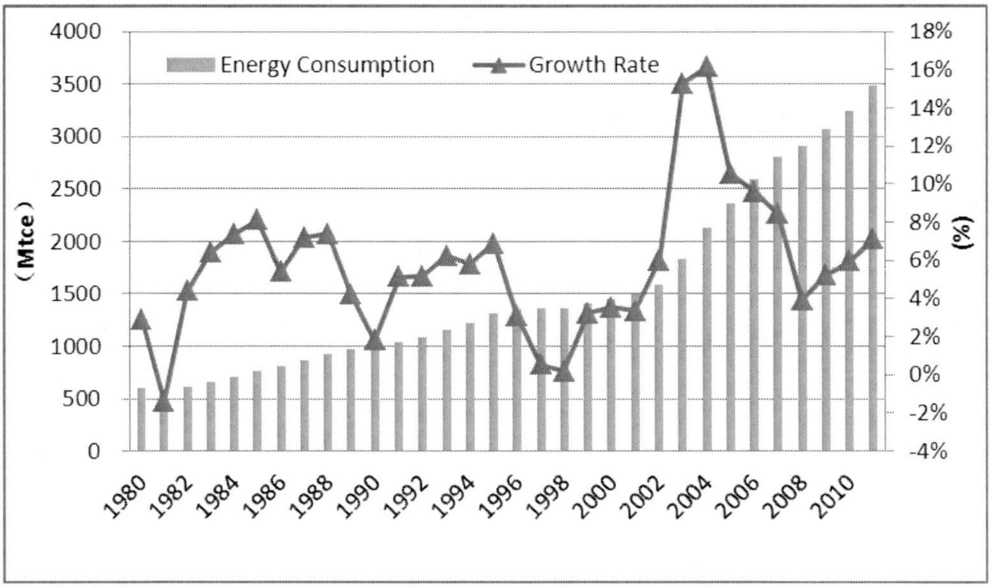

Figure 1-2. Energy consumption and growth rate in China, 1980-2011.

Economic growth is accompanied by rapid energy consumption. Various empirical studies have revealed strong evidences that Chinese economy is energy-dependent [2]. As is indicated in Figure 1-2, growth of energy consumption is almost synchronous with China's economic growth during 1980-2011. It is worth noticing that China has made huge progress in energy efficiency in the first two decades of reform-and-opening-up. From 1980 to 2000, China managed to fuel fourfold GDP growth with just double energy consumption growth. However, energy consumption experienced rapid growth in the early of 21st century when

China entered a new stage of rapid industrialization and urbanization. The annual growth of energy consumption was 10.95% during 2002-2007, which is more than double the growth rate during 1980-2000. The growth slowed down after 2006 and in 2010 the growth rate was 6.0%, thanks to the active efforts of the government to cutting down energy intensity of the economy [3].

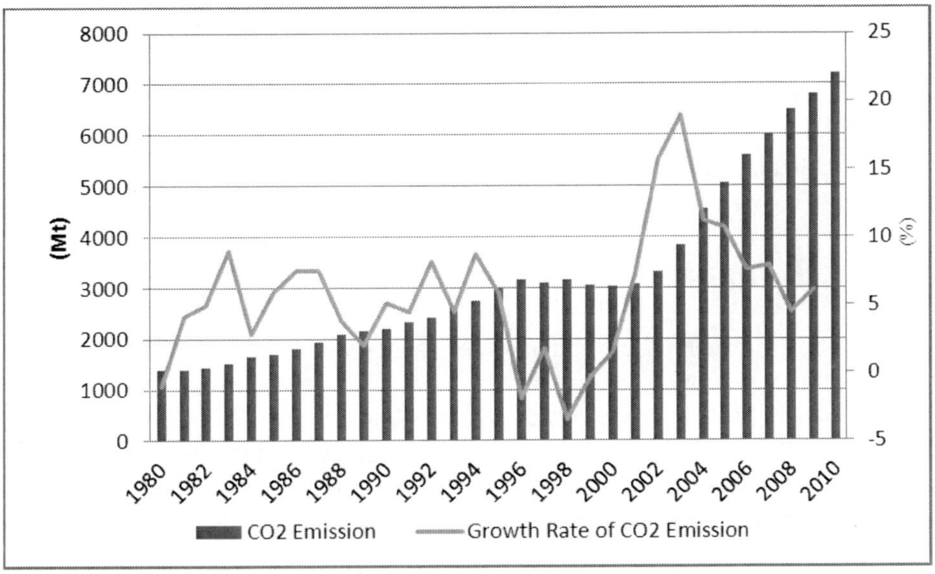

Figure 1-3. CO_2 emissions and growth rate in China, 1980-2010.

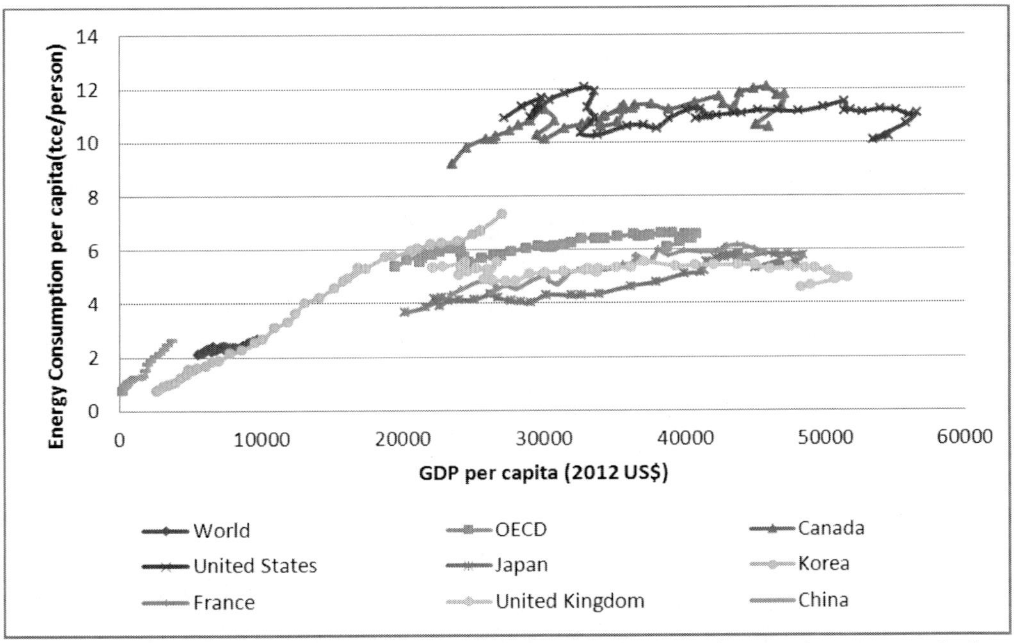

Note: in IEA the US$ is in 2005-year constant price. For consistence of comparison, we recalculate the data into 2012 US$ according to exchange rate of US$-RMB in 2005 and 2012.

Figure 1-4. Per capita energy consumption vs. GDP per capita in the world, 1971-2010.

An inconvenient companion of increasing energy consumption is the enlarging CO_2 emissions in China (Figure1-3) [4]. In 2009 China surpassed the United States of America and became the largest CO_2 emitter in the world [4]. Considering the fact that income level of Chinese resident has just reached the upper middle income and vast growth potential of Chinese economy, the global society has serious concern on China's future emissions trajectory. This concern has aroused widespread discussions and the International Energy Agency (IEA) published the2009 edition of World Energy Outlook, which set out an aggressive "450 Scenario'' by 2030 with China-specific policy assumptions and outlook [5]. As a response to the global concern, President Hu Jintao declared a target of reducing the CO_2 intensity of Chinese economy by 40-45% in 2020 on the baseline level of 2005 in the 2009 United Nations Climate Change Conference (COP15).

1.2. Comparison with Other Countries

Figure 1-4 and 1-5 present the comparison of energy consumption and CO_2 emissions per capita along with the growth of income per capita in China and other countries. It is evident that with an absolute low income level (3725US$ in 2010), China's per capita energy consumption is still low (1.7 toe in 2010), relative to that of industrialized countries (about 4.5 toe for the OECD average in 2010). But an inconvenient truth is that China's per capita CO_2 emissions (5.7 tonnes in 2010) are approaching the levels of several high income countries including the United Kingdom, France and Japan (6-8 tonnes in 2010) quickly. A natural implication is that if the current trend kept as usual, China's per capita CO_2 emissions would soon reach the average level of high income countries while its income level was still much lower.

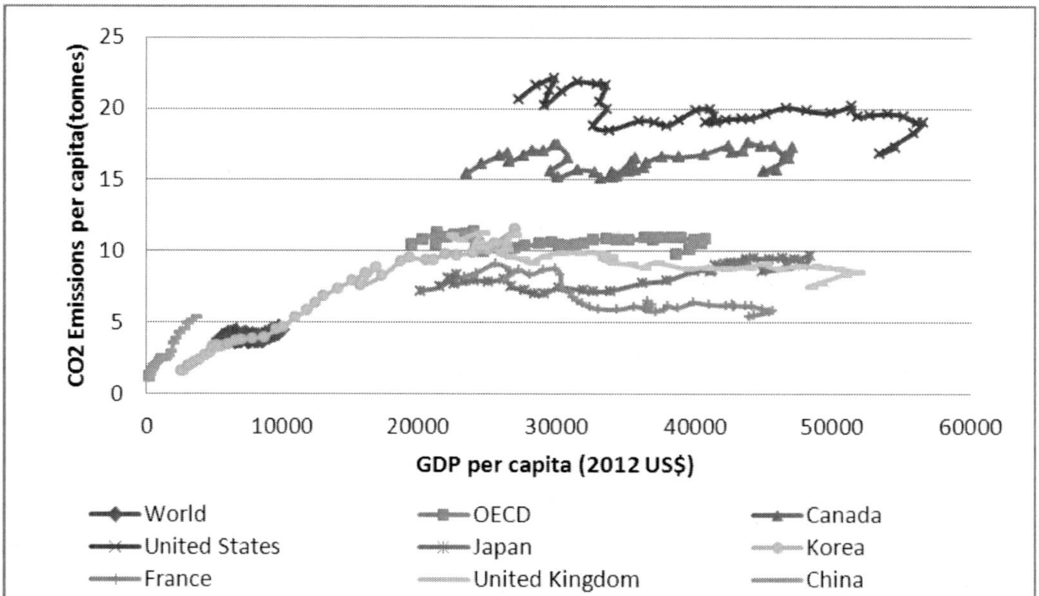

Figure 1-5. Per Capita CO_2 emissions vs. GDP per capita in the world, 1971-2010.

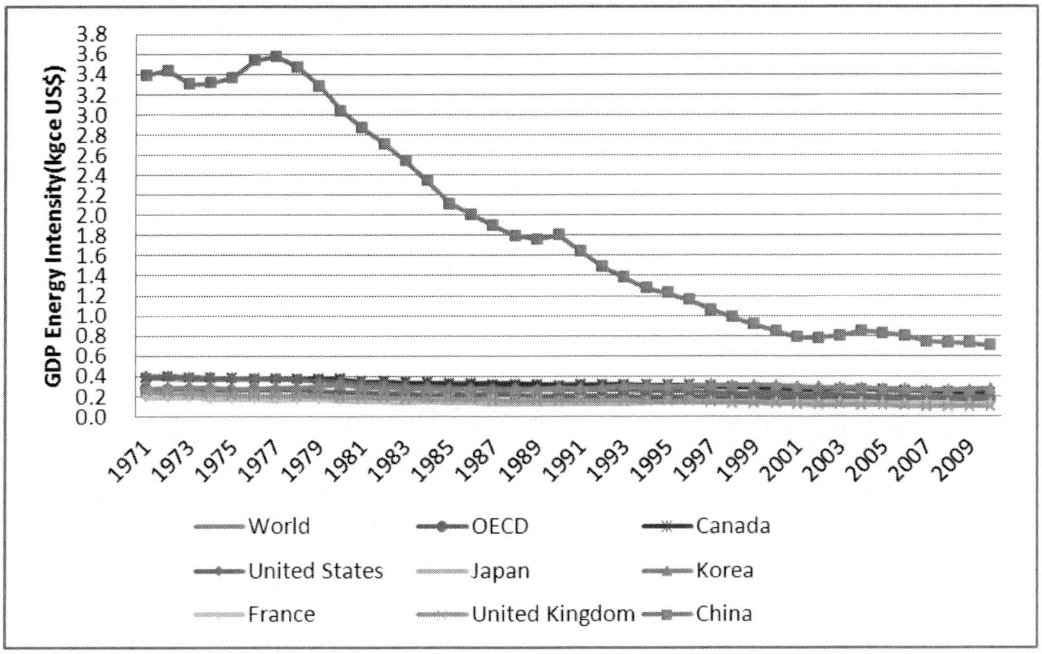

Figure 1-6. Energy intensity of the economy in the world, 1971-2010.

1.3. Underlying Factors

Two interrelated factors lead to China's asymmetric high CO_2 emissions level. The first is China's poor energy efficiency. Figure 1-6 presents a comparison of energy efficiency in China and other countries [4]. It is evident that the level of China's energy efficiency is not only much lower than that of industrialized countries but also significantly lower than these of other developing countries.

Fossil energy (especially coal) based primary energy structure is the second factor. China's primary energy supply is dominated by coal (Figure1-7) [1]. Coal has long accounted for about 70% of primary energy consumption. The share of oil rose from 16.6% in 1990 to 22.3% in 2002. Due to supply limit and price hike, its share dropped somewhat to 17.9% in 2009. Gas held small share in primary energy supply but its share rose stably from 1.8% in 1995 to 4.4% in 2010. Non-fossil energy, including hydropower, nuclear power and wind power, though grew stably due to China's clean energy development strategy, but the absolute share is still marginal. Meanwhile, heavy reliance on contributes partly to low energy efficiency in China. In the future, fast economic growth prospect and increasing energy resource-environment-ecology constraints will compel Chinese Government to pay more attention to renewable energy development.

China has been the world's fastest growing market for renewable energy in recent years. Its newly increased installed renewable energy capacity was 133 GW in 2011, with the top five-year growth rate of 92% since 2006 [6]. It is also the world's number one in terms of installed capacities of hydropower, wind energy and solar water heat. Figure 1-8 shows installed capacities of renewable generation in China from 1990 to 2011. It is clear that small hydro has been developed steadily since 1990 while the development of wind energy and solar PV accelerated more recently especially after the promulgation of the Renewable Energy Law in 2006. Particularly, the implementation of a national feed-in tariff for solar PV in 2011 has almost tripled its installed capacity from 0.81 GW to 3.1 GW in 1 year. Other renewable energy technologies such as geothermal and wave power generation are at the demonstration phase with minimal installed capacities in operation.

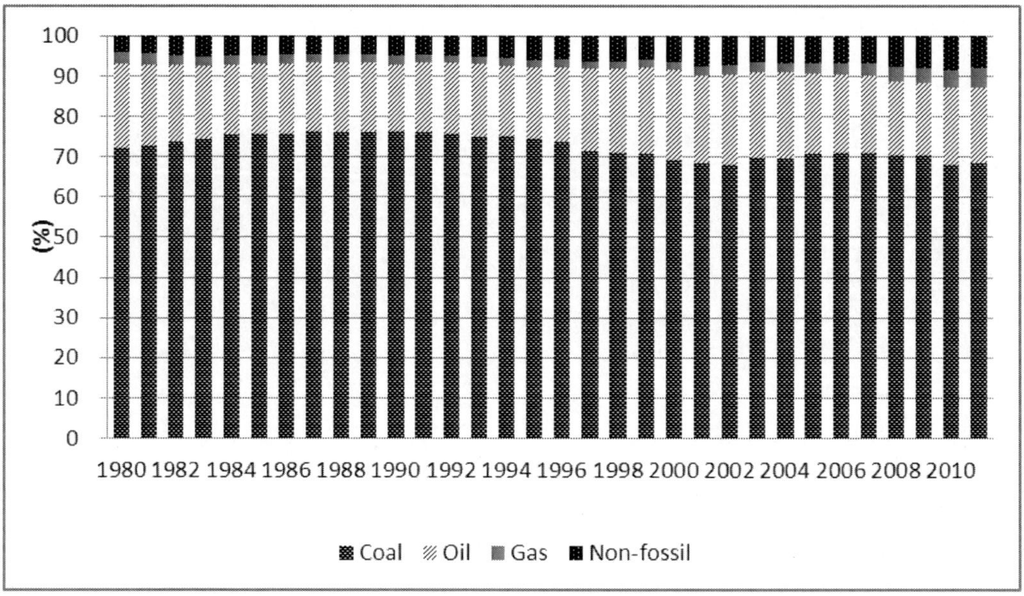

Figure 1-7. Primary energy consumption structure in China, 1980-2011.

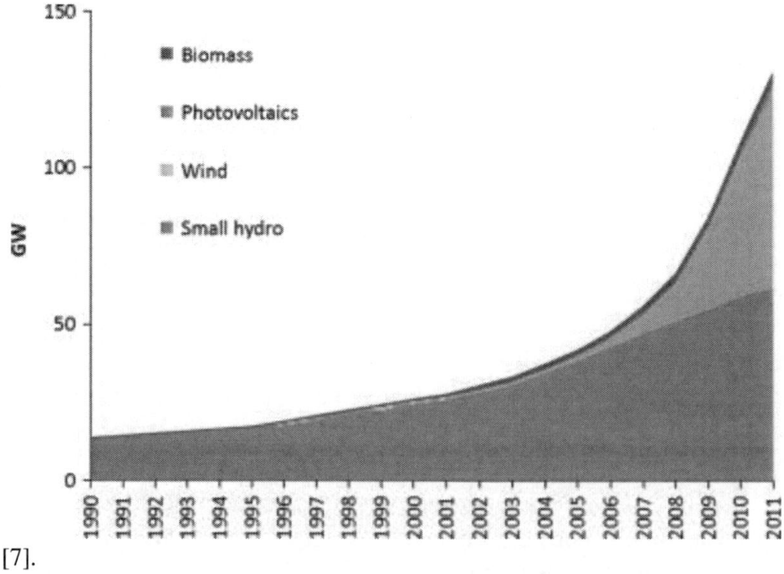

Note: Source [7].

Figure 1-8. Growth of renewable power generation capacity in China, 1980-2011.

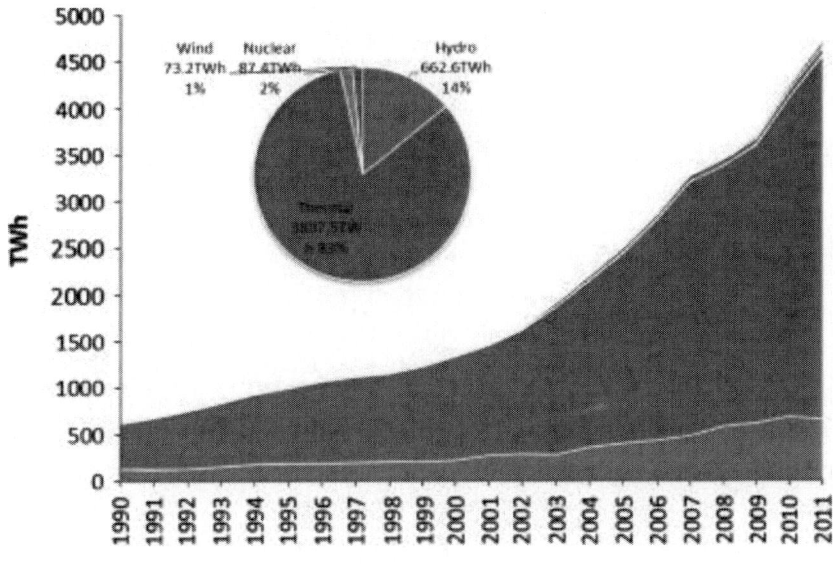

Figure 1-9. Power generation structure in China, 1980-2011.

Despite a great amount of accumulated installed capacities in renewables, the share of renewable (excluding hydropower) in electricity generation was merely 1.2% in 2010. As seen in Figure 1-9, the electricity generation mix of China is still dominated by thermal power plants, which produced 83% of electricity in 2011 [1, 7]. From Figure 1-10, it is evident that there is big gap between China and several major developers of renewable power generation in the world [8-9].

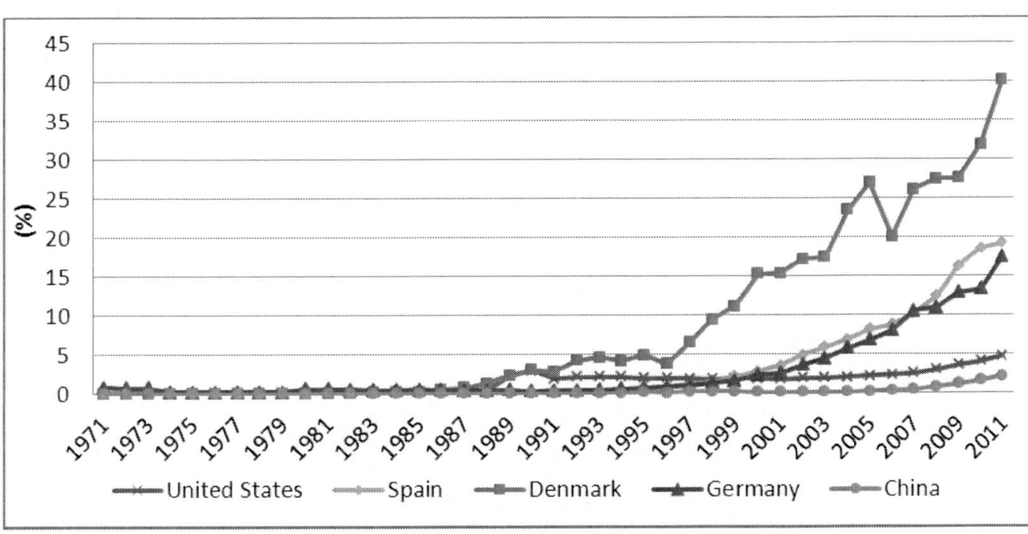

Note: Include geothermal, solar, tide, wind, biomass and biofuel energy, without hydroelectric.
Note: Source [8-9].

Figure 1-10. The proportion of renewable power generation in some countries.

2. Renewable Energy Resource, Legistlaiton and Planning in China

2.1. Renewable Energy Resource in China

2.1.1. Hydropower

China has abundant hydropower resources. Based on the 4th national survey of hydropower resources conducted in November of 2005, the gross theoretic hydropower reserve and annual average power generation potential is estimated at 694 GW and 6083 TW h respectively. The technically exploitable capacity is illustrated as 542 GW and the corresponding annual average power generation is 2474 TW h which ranks first in the world, while economically exploitable capacity and power generation are 402 GW and 1753 TWh respectively [10]. National hydropower resources also include the ample resources of small-sized hydropower stations (which refer to stations with an installed capacity below 50 MW). The technological potential of small hydropower resources in China is 128 GW and the power generation is 450 TWh.

However, the distribution of hydro resources in China is uneven. Southwest China which includes four provinces Sichuan, Tibet, Yunnan, Guizhou and Chongqing (municipality directly under the central government) has the most hydropower resources in this country, followed by Mid-South China, and lastly North China (Figure1-11).

2.1.2. Wind power

Throughout China's vast land mass and long coastline there is a rich resource of wind energy with great development potential (Figure 1-12). According to the last national investigation, the technically exploitable land area (with a wind power intensity of over 150 W/m^2) is approximately 200,000 km^2. Taking a ratio of 3-5 MW/km^2, the technically and economically exploitable wind power resource would be 600-1000 GW. According to the resource survey in coastal regions, there is also 157,000 km^2 of coastal areas around China with a water depth between 0 and 20 m. The installed capacity of offshore wind power could reach 100–200 GW if wind power can be realized at a density of 5 MW/km^2 over 10–20% of the sea. In summary, there is a huge potential for wind power in China at around 800-1200 GW [11].

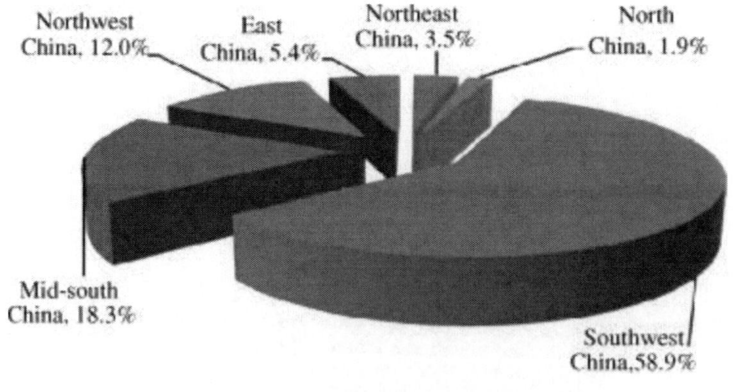

Figure 1-11. Regional distribution of hydropower resources in China.

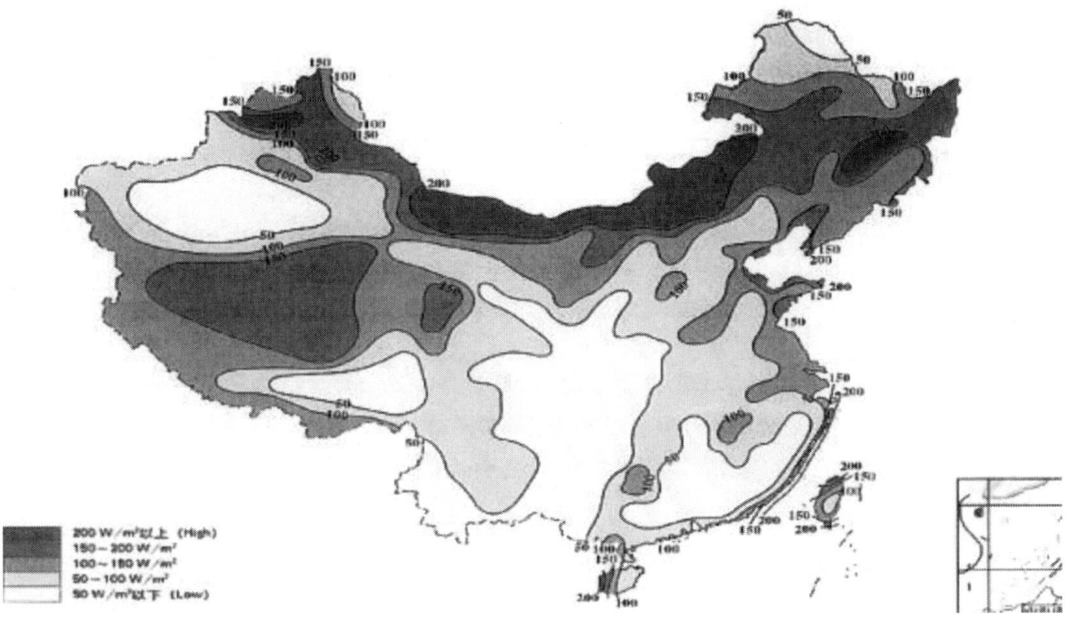

Figure 1-12. Regional distribution of wind power resources in China.

Although there are excellent wind resources across the country, the geographical distribution of wind resource does not fit well with the power load profile. In economically developed south-east coastal regions with large power load, there are fewer wind resources; while in northern regions (north-west and north-east) regions with massive wind energy potential, the demand for power is smaller. This creates difficulties for the economic development of wind power in most of the areas with abundant wind resources, which are far away from the power load centers. But in some areas such as Jiangsu, Fujian, Shandong and Guangdong, there are vast offshore resources located close to the centers of power demand. With its gradual technological and economic maturity, offshore wind energy will play an important role in the power supply for these areas.

2.1.3. Solar power

China lies in the northeastern part of East Asia between 4°and 53°North latitude and 73–135°East longitude covering an area of 9.6 million km^2. More than two-third of the country receives an annual radiation of more than 5000 MJ/m^2 and more than 2000 hours of sunshine [12]. The areas with large amount of radiation include among others Tibet, Qinghai, Xinjiang, southern Inner Mongolia, Shanxi, northern Shaanxi, Hebei, Shandong, Liaoning, western Jilin, middle and southwestern Yunnan, southeastern Guangdong, southeastern Fujian, eastern and western Hainan Island and southern western Taiwan (Figure1-13). The Qinghai-Tibet Tableland, in particular, receives the most radiation.

2.1.4. Biomass energy

The employable biomass energy in China mainly includes crop stalks, firewood, manure, domestic garbage, industrial organic waste residue and wastewater. It is estimated that the total exploitable biomass energy in China is around 700 million tce. Among the exploitable biomass energy resources, 40% of the crop stalks are used as feedstuff, fertilizers and industrial raw materials and the rest of 60% (210 million tce) are used for energy. Firewood is mainly used as fuel, but about 40% of the remnant forestry (30 million tce) has not yet to be utilized. A small proportion of human or animal manure is used as fertilizer, while the major part (60 million tce) is left as the principal source of pollution in rural areas. At least 80% of the industrial organic waste residue can be redeployed, approximating 70 million tce. Finally at least 80 million tce can be generated from domestic garbage. Consequently, China's theoretical biomass energy that can be used as energy could reach at least 500 million tce [13].

2.1.5. Ocean energy

Ocean energy resources include tidal energy, wave energy, oceanic flow energy, temperature difference energy, and salt difference energy, among others. According to rough estimates, the theoretical reserve of tidal energy is up to 190 GW, among which 90% are distributed in Zhejiang and Fujian province. The average theoretical reserve of wave energy resources is 12.9 GW, among which most are located in Zhejiang, Guangdong, Fujian and coastal areas in Shandong. The theoretical average power in oceanic flow resources is 139 GW, among which half are distributed in Zhejiang. The theoretical reserve of ocean temperature difference energy resources is around $(1.19\text{-}1.33)\times10^{19}$ KJ, the technically exploitable (heat efficiency at 7%) part is about $(8.33\text{-}9.31)\times10^{17}$ KJ, with a theoretical potential of 1320-1480 GW. The reserve of salt difference energy resources is about 3.9×10^{15} KJ, with a theoretical potential of 125 GW [14].

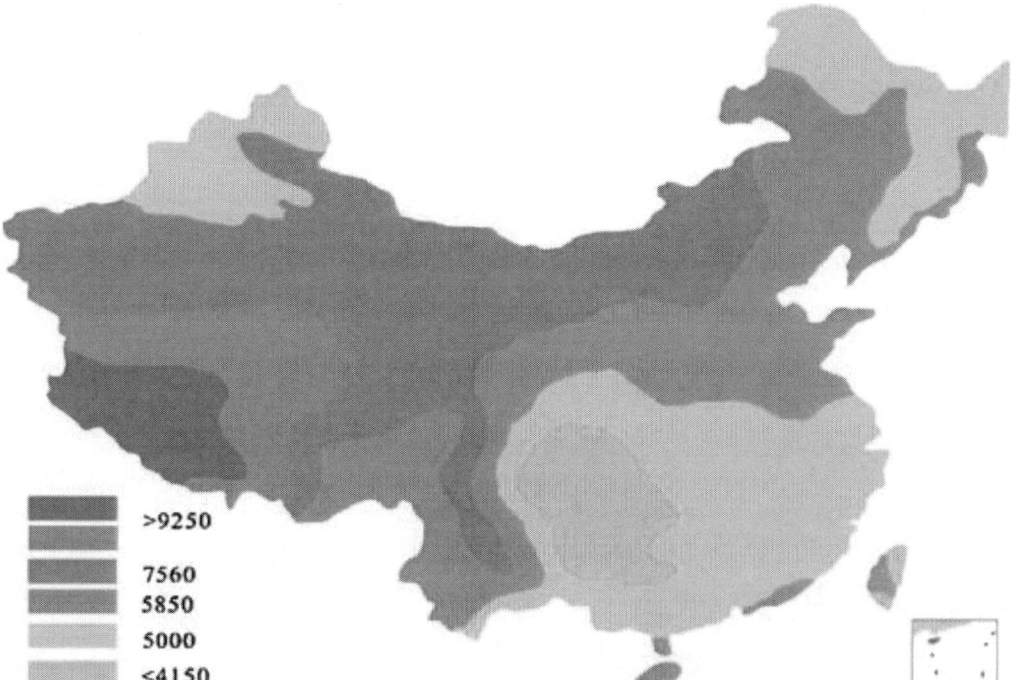

Figure 1-13. Regional distribution of solar power resources in China.

2.1.6. Geothermal energy

China's geothermal energy is mainly located at the circum-Pacific tropical zone and Himalaya-Mediterranean tropical zone. More than 3200 geothermal spots have been found in China, with an annual naturally heat release of 1.04×10^{17} kJ, i.e., 3.56 billion tce, among which 80% are geothermal hot land below 100 °C.

2.1.7. Overall renewable energy resource in China

According to the above analysis, we compile table1-1 for a summary of renewable energy resources in China [15-16]. Note that the annual generation potential of hydropower, wind power, biomass and solar power is estimated according to the technically and economically exploitable resources. But due to the unavailability of the data for geothermal energy and ocean energy, their potential estimate is based on theoretical resource reserve, which should be amended when more detailed resource survey are conducted in the future.

2.2. China's Legislation for Renewable Energy

China has made impressive strides in recent years on renewable energy legislation so as to increase its use of renewable energy. This effort was kick-started with the passage of the *Renewable Energy Law of the People's Republic of China* (*Renewable Energy Law*). On February 28, 2005, the Standing Committee of the National People's Congress ("NPC") promulgated the *Renewable Energy Law*, which became effective on January 1, 2006 [17]. As stated in the law, the purpose of the Renewable Energy Law is to promote the development and utilization of renewable energy, increase energy supplies, improve the energy structure, guarantee energy security, protect the environment and realize economically and socially sustainable development. The law stated that renewable energy development and utilization would be treated as a priority area for energy development, and defined renewable energy to include wind, solar, hydro, biomass, geothermal, ocean and other non-fossil energy. Subsequent regulations have clarified that biomass power includes agricultural and forest waste-generated power, as well as waste-to-energy power and sludge gas power projects.

The *Renewable Energy Law* created for the first time a national framework for promoting renewable energy development in China, by establishing four key mechanisms to promote the growth of renewable energy in China:

(1) *a national renewable energy target and central and local renewable energy development and utilization planning*; (2)*a mandatory connection and purchase policy* by which grid companies are required to sign an agreement with renewable electricity generators in their jurisdiction to purchase all of the electricity generated from the generators, and provide grid connection services;(3) an *on-grid electricity price* for renewables, akin to a *national feed-in tariff system*, which pays renewable electricity generators a fixed, additional amount for each kilo-watt hour of electricity generated, above the wholesale electricity price for desulfurized coal-fired power; and (4) *a cost-sharing mechanism*, funded by a surcharge on electricity sales, to pay for feed-in tariffs, grid connection projects, and independent, public renewable energy grids, as well as a *Renewable Energy Development Special Fund* to cover

activities such as science and technology research for renewables, standard setting, pilot projects, rural utilization of renewables, and renewable resource assessments.

Each of these policy mechanisms works together with and complements the others; conversely, many of the challenges China has and continues to face in developing and utilizing renewables requires examining each of the mechanisms to see how they can be improved to bolster each other and function together to provide a stronger policy framework for renewables development [18].

In the first few years following implementation of the law, China's installation of renewables, wind power capacity in particular, grew at a spectacular rate. Wind power doubled in capacity each year during the four year period 2006-2009, and wind has grown from a mere 1.3 GW at the end of 2005 to 62.4 GW (grid and non-grid connected turbines) by the end of 2011, leading China to surpass the US as the country with the great installed wind capacity in the world. However, China's rapid growth in wind power has been accompanied by significant growing pains, which have exposed areas in which the renewable policy framework can be improved, including: long delays in connecting wind farms to the grid, due to a lack of coordination between wind farm and grid development planning, as well as a lack of incentives for grid companies to connect to renewables generators (about 25-30% of China's erected wind turbines are not connected to the grid at any one time, although most eventually are connected); the lower conversion efficiency of domestic turbines compared to international manufacturers; and serious curtailment of wind-generated electricity due to concerns that wind-generated electricity might destabilize the grid [18-20].

In short, even though the Renewable Energy Law had specified that grid companies were required to connect and purchase all electricity from renewable sources, this often did not happen in practice. Renewables development has focused on the construction of wind farms and solar farms, with much less attention paid to the actual generation and utilization of electricity from such projects. Renewables development also suffered due to delays in payments to renewable electricity generators for FIT payments or to compensate for grid connection projects.

Aware of the challenges affecting renewables development, on December 26, 2009, the NPC passed amendments to the Renewable Energy Law, which took effect April 1, 2010, and included several significant changes meant to improve the renewable energy framework in China, including: adding measures intended to improve implementation of the mandatory connection and purchase policy, such as a renewable power quota system providing targets based on generation rather than installed capacity, a priority dispatch system that favors renewable electricity, and technical standards for interconnection to the grid; streamlining the renewable energy fund, in order to speed up payments for renewables incentives such as the FITs; and increasing central government oversight of provincial and local renewable energy development planning [21].

The 2009 amendments to the Renewable Energy Law and the regulations that have or are being promulgated in conjunction with the amendments show promise as solutions to address some of the challenges to renewables development in China.

Table 1-1. Renewable Energy Sources in China

Type	Resource reserve	Annual generation potential
Hydropower	Theoretical reserve: 694 GW; technically exploitable: 542 GW (including 128 GW small hydropower); economically exploitable: 402 GW	1800 TWh
Wind power	Theoretical reserve: 2600 GW onshore resource and 500 GW offshore resource; Technically and economically exploitable: 1000 GW	2000 TWh
biomass	600-700 million tce/year including biofuel	1500-1750 TWh
Solar power	2/3 of the country receive an annual radiation of more than 5020 MJ/m^2 and more than 2000 h sunshine; theoretical reserve: 1700 billion tce/year	>128000 TWh
Ocean energy	Theoretically exploitable resource: 1600-1800 GW; Technically and economically exploitable resource: unclear	1200-1500 TWh
Geothermal energy	Theoretical resource reserve: 3.56 billion tce; Technically and economically exploitable resource: unclear	8000-10000 TWh

2.3. Government Entities Related to Renewable Energy in China

On national level, there are various ministries responsible for energy issues.

- The National Energy Administration (NEA) is situated within the National Development and Reform Commission (NDRC), China's top economic planning agency, and is responsible for development and policy planning with respect to energy, including planning for renewable energy development.
- The NDRC Pricing Bureau is responsible for setting prices for energy, among other economic goods, including setting the on-grid price for wind generated power, biomass power, solar PV power, etc.
- The State Electricity Regulatory Commission (SERC) is a government agency responsible for regulating China's electricity industry, including the integration of renewable electricity into the grid. In March of 2013, SERC was dismantled and its functions are merged into NEA.
- The NDRC is in charge of collecting funds from the renewable energy surcharge and distributing them to grid companies, and more importantly, is also responsible for the formulation of industry policy relevant to renewable energy and the approval of renewable energy projects.
- The Ministry of Finance (MOF) is in charge of the Renewable Energy Development Special Fund and its supported programs.
- The Ministry of Science and Technology (MIST) is responsible for the R&D of renewable energy technology.

- The Ministry of Industry and Information Technology (MOIIT) is responsible for industry policy, especially industry policy for the renewable energy technology and equipment manufacturing, and energy conservation in industry.

Besides, Ministry of Agriculture and State Forestry Bureau are also relevant with biomass development, especially in rural areas in China. Because there is not a unified Ministry of Energy, like DOE in the USA in charge of unified energy strategy, energy planning and energy policy, historically the formulation of energy policy and its implementation in China is based on the perspective of individual ministry, not in a holistic way. Therefore, lack of foresight, lack of synergy or even conflicts among the policies from different ministries, and lack of consistency are among the main shortcomings of energy regulations in China. As a response to it, in 2010 the National Energy Committee (NEC) was established to function as the top authority on energy issue in China. Premier of the State Council is the director of the committee while the ministers relevant to energy issues are its members. The Office of NEC is situated in NEA. In this way, NEC is in charge of coordination among ministries on energy strategy and planning while NEA is the implementation body of NEC.

On provincial level, local government also has important roles in energy issue. For example, the wind power projects below 50 MW are approved by provincial governments according to the existing delineation of power between central and provincial governments. However, because of approval procedure of central government (by NDRC) is tedious and stricter, provincial governments are motivated to decompose a big wind power project into several small projects and approve them under their jurisdiction. In turn, the consequence is that the planning for renewable is inconsistent between national and provincial levels. Furthermore, though provincial governments are interested in the development of renewable energy projects, they may have complicated sentiment on the utilization of renewables. For example, when formulating the annual power generation planning, provincial government prefers coal power over wind power, because coal power is reliable and because coal power can contribute more tax revenue [22].

2.4. China's Overall Planning for Renewable Energy

The 2005 *Renewable Energy Law* tasked the State Council's energy department (the energy body under the National Development and Reform Commission (NDRC))4 with devising mid- and long-term national targets for renewable energy production (Art. 7), as well as a national renewable energy development and utilization plan (Art. 8). Accordingly, in 2007 the State Council released the *Mid- and Long-Term Development Plan for Renewable Energy*, which set targets for meeting 10% of China's primary energy consumption from renewable energy sources, including hydropower, wind, and solar by 2010 and 15% by 2020. The *Mid- and Long-Term Development Plan for Renewable Energy* also established a "Mandatory Market Share" (MMS) requirement for grid companies and generators for non-hydro renewable power: grid companies were to achieve 1% of their total power generation from non-hydro renewable power by 2010 and 3% by 2020, and investors (i.e., generators) with authority to install over 5 GW of power capacity were required to achieve 3% of their total installed capacity from non-hydro renewable power by 2010, and 8% by 2020. The focus on non-hydro renewable power targets was reasonable given that China had already developed a large amount of hydropower (14.5% of total electricity generation in 2007), while wind, solar, biomass power and other renewable power technologies all were far below 1% of generation in 2007.

The Mid- and Long-Term Development Plan for Renewable Energy also set specific installed capacity targets for 2010 and 2020 for various renewable power technologies, including hydropower, wind and solar (see Table 1-2). The 2020 installed capacity target for wind power, 30 GW, was already met by 2010, highlighting the challenges in setting targets for rapidly expanding renewable technologies.

In 2008, NDRC released the *Renewable Energy Plan for the Eleventh Five-Year Period*, which reiterated the target of reaching 10% of energy consumption from renewable sources by 2010, the final year of the Eleventh Five-Year Plan. The 11th Five-Year *Renewable Energy Plan* also set out target indicators for the installed capacity and generation to be met by various technologies, including hydro, wind, PV, and biomass power.

Table 1-2. Renewable Energy Planning in China

Renewable energy	Target of 2010	Actual in 2010	Target of 2015	Target of 2020
Overall target				
Non-fossil fuel share in primary energy	10%	9.4%	11.4%	15%
Renewable energy share in electricity supply (excluding hydro)	1%	1.2%		3%
Electricity generation (GW)				
Hydro	190	272.6	290	300
-including small hydro	50	59.2		75
Wind	5	44.7	100	200
Solar PV	0.3	0.8	21	50
Wave				0.1
Biomass	5.5	4	13	30
Biogas supply (million m^3)				
Total	19,000	7455 (2009)		44,000
From household	15,000	2000 (2009)		30,000
From large livestock farm	3000	455 (2009)		
From industrial waste	1000	5000 (2009)		
Biofuel (million tons)				
Bioethanol	3	1.68		10
Biodiesel	0.2	0.2		2
Solid biofuels	1			50
Heat supply				
Solar water heat (million m^2)	150	168		300
Geothermal utilization (TWh)	27.78			97.69

In 2009, when China committed to the 2020 climate and clean energy goals ahead of the Copenhagen climate conference, the 2020 target of obtaining 15% of all of its energy from non-fossil energy sources was expanded to encompass all non-fossil energy, i.e., renewables as well as nuclear power. These climate and non-fossil energy goals were part of an evolving energy saving and clean energy policy that also included goals in the 11thFive Year Plan (FYP) (2006-2010) to reduce energy intensity (the amount of energy consumed per unit GDP) by 20%, and in the 12th FYP (2011-2015) to reduce energy intensity by a further 16% and to reduce carbon intensity (the amount of carbon emitted per unit GDP) by 17%. The 12th FYP also set a mandatory target to obtain 11.4% of primary energy consumption from non-fossil sources by 2015.

3. Institutional Context for Renewable Energy in China

3.1. The Social-Political Context for Electric Power Sector in China

Because most of the renewable energy is utilized through power generation, it is important to understand the social-political context for the electric power sector in China [23]. The current power regime in China for meeting industrial process, lighting, heating and energy-related services may be characterized as a centralized system. Electricity is centrally generated, largely from coal, hydropower, and a small but growing amount of nuclear and renewable sources; it is delivered to businesses and homes through large scale transmission and distribution networks, before being used to provide power, lighting, heating and services with the aid of end-use technologies and the buildings infrastructure. The absolute low starting point, the vast growth in power demand incurred by rapid industrialization of Chinese economy and the rise of China as the World Workshop, as well as the increasing demand from households as a result of enhanced living standards and the popularization of home appliances, have made rapid growth of the power supply capability a policy priority in China. The strategic importance of energy to enabling economic activity and well-being means that the system is the subject of intense policy activity, which focuses on ensuring secure and affordable supplies, and other social objectives. The meta "socio-technical landscape" of electrification in China can be characterized as:

- a dependency upon fossil-fuel-based energy supply (mainly coal and rising supply capability of gas) and large-scale electricity generation technologies,
- a coal extraction, processing and transportation infrastructure always unable to catch up with increasing demand,
- province-based power grids and a growing interconnected national grid to ease the geographic imbalance between supply and demand,
- historically vertical integration monopoly just being restructured into two grid operators and a set of SOE generators who are regulated by government bodies,
- a weak but expanding rail and road infrastructure,
- a traditionally centralized administration system, with initiated but rather fruitless market reform.

Key processes that influence or 'drive' the power regime at the landscape level in China include:

- the increasing power demand from the industrial sector because of the on-going industrialization process and a big gap in energy efficiency compared to the rest of the world,
- the increasing power demand from households because of increasing income and rapid urbanization,
- a huge population base of 1340 million, which is still growing,
- the increasing pressure on primary energy supply, environment and ecology because of the heavy reliance on coal,
- global concern of GHG reductions and increasing pressure on China's "measureable, reportable and verifiable " obligation to cut GHG emissions,
- the Chinese Government's commitments to reduce emissions, and reduce GDP carbon intensity by 40-45% in 2020 compared to the 2005 level and to promote clean energy sources, increasing the share of non-fossil over primary energy by 15% in 2020,
- concerns over security of primary energy supply,
- external factors leading to high and/or volatile oil and gas prices and related concerns over energy affordability and fuel poverty, as well as physical disruption of external supplies (war, terrorism, foreign governments limiting supply, etc.),
- the gradual transition of China's economy from centralized planning to a more market-based economy that will gradually blend into the global economic system,
- and the international efforts to deliver renewable energy technologies.

3.2. China's Divergences from the "Best Practice" of Renewable Energy Deployment

The "best practice" to deploy renewable energy includes: (1) policies that eliminate economic barriers to renewables by leveling the playing field of renewables vis-à-vis fossil fuels, as well as by implementing support mechanisms that compensate for high costs, limited access to finance, and insufficient demand; and (2) institutions that eliminate non-economic barriers by ensuring good governance on the part of the State and corporate competition [24].

As shown in Table 1-3, many of the "best practice" recommendations constitute ideals, unlikely to be implemented in full in any real context. Some items of the "best practice" model are even impossible to account for (for instance, how precisely can it be established whether an externality has been fully internalized?). Nevertheless, we still find it compelling to explain how and why China's regulations diverge from the model.

China's policies and institutions fit into the "best practice" model only partially and imperfectly. More concretely, China's policies diverge from best practice insofar as: negative externalities of fossil fuels are not compensated for (as with a coal tax); regulations do not incentivize feeding power into the grid, but instead focus on installing capacity (China's renewable portfolio standard does not refer to the actual power fed into the grid but to the installed capacity; and the tender system for wind that prevailed until 2009 had no provisions to ensure generation and transmission); remuneration levels are low and duration of tariffs is short, be they tariffs set in tenders, in local licenses, or through feed in tariffs (FITs); regulations do not include enough provisions for the reduction of tariffs over time, necessary for the promotion of cost-reducing innovations; and power purchase agreements (PPAs) do not ensure connection (Table 1-3).

Meanwhile, concrete divergences from best practice in institutions include the following: general legal insecurity; complex and lengthy red-tape; unpredictable policy instruments (insufficient stability and transparency); insufficient competition in generation due to market concentration, a high market share remaining in public hands, and limits to foreign presence; and restrictions to innovation in manufacturing brought about by barriers to external trade and to foreign investment [24].

Table 1-3. China's divergences from the "best practice" of renewable energy deployment

	Policies and institutions for renewables in the "best practice" model	**China's divergences from "best practice"**
Policies to overcome economic barriers	Elimination of coal subsidies	Negative externalities not fully compensated for
	Compensation for the negative externalities of fossil fuels (pollution, etc.)	
	Remuneration for the positive externalities of renewables	
	Compensation for high initial costs (mandated market policies): quantity-based and price based schemes	Regulations focus on installed capacity rather than power generation
		Remuneration levels are low, and duration of tariffs is short
	Increased access to capital: fiscal and financial aids	Regulations do not include enough provisions for reduction of tariffs
	Ensuring sufficient demand (PPAs)	PPAs do not ensure connection
Institutions to overcome non-economic barriers	General legal security	General insecurity and uncertainties
	Capable bureaucracy: coordination and cutting of red-tape	Incomplete coordination, and complex and lengthy red-tape
	Quality of regulations in renewables: specific, legally binding targets, and predictable instruments	Targets not compulsory, and instruments lacking stability and transparency
	Competition and technology-friendly policies in generation: unbundling, absence of oligopolies, openness to FDI	Limits to competition in generation (market concentration, public ownership, and barriers to foreign entry)
	Competition and technology-friendly policies in manufacturing: openness to external trade and FDI	Limits to innovation (barriers to foreign trade and foreign entry)

3.3. Institutional Context for Energy Issue in China

3.2.1. Developmentalism

As in the paradigmatic cases of Japan or South Korea, in China developmentalism entails that: (1) economic policy has development goals (as in other countries, the idea of "development" in China has evolved with time; more particularly, policy stressed growth targets in the 1990s, and – as we will point out – in the current century it pursues a broader concept of development, inclusive of social and environmental parameters); (2) development is deemed as necessary for political legitimacy and stability; and (3) development is to be achieved by means of the State's involvement in the mobilization and allocation of resources.

First, the Chinese government is widely recognized as having developmental goals, in the present century with an emphasis on socially equitable and environmentally sustainable growth – an emphasis embedded in the idea of Scientific Development. It should be stressed that China shares with prototypical developmental states an emphasis on development goals attached to a somewhat lesser emphasis on rules: concreteness and transparency of regulations are not necessary for development; and ideology can be set aside when deciding regulation, opening the door to pragmatism, flexibility, and eclecticism in the choice of policies and institutions.

Bringing the developmental state to electricity and renewables, there are very diverse development goals – non-renewable-energy goals – embedded in China's decisions regarding renewables. Goals include energy security (limiting oil imports, avoiding black-outs), socio-economic development (developing local industry, providing employment, lessening rural-urban inequalities and consequent migration, etc.); and environmental protection (diminishing local pollution, as well as emissions of greenhouse gases). Also, the emphasis on deploying capacity (and not on feeding power to the grid) favors local industry, provided that local companies are able to install capacity but are less capable than foreign generators to efficiently run capacity. Finally, the limited remuneration and duration of support schemes also helps local companies, given that these are able to operate with lower profit margins and with higher uncertainty than their foreign counterparts. An emphasis on setting goals, instead of an emphasis on making regulations concrete and transparent, is also observable in Chinese policies and institutions for renewables.

Second, China's developmentalism is frequently seen as the means to preserve its political regime. For the case of electricity, electric sector reform was not an embrace of competitive market models, but the "creative, dynamic response to a set of technical and economic constraints on the one hand, and the political imperative to stay in power on the other. This logic of reform motivates the strategic decision to increase electricity production in order to meet current demand and fuel future economic growth. Such growth, in turn, is part of a larger effort by the party-state to maintain legitimacy by channeling potential citizenship demands into consumption and thus pacifying newly middle-class consumers. Similarly, it is arguable that if Chinese policies and institutions for renewable do not fit into "best practice", it is because these are not an advancement toward the perfection of electricity markets and the creation of market-friendly institutions. Rather, they are the necessary response to diverse development needs that, if unattended, could lead to a loss of legitimacy of China's political regime.

Third, the policy and institutional instruments to achieve developmental goals are close to those of developmental East Asia. Essentially, these include a wide array of non-market-distorting instruments, as well as instruments that do distort resource allocation. In other words, the role of the State in China is not one of creating and perfecting markets, or of ensuring that the proper market institutions are in place, but rather to control these in search of the aforementioned developmental goals. Involvement of the State in the allocation of resources is exerted via a range of mechanisms that extend from indicative planning to industrial policy and direct ownership of companies. Indicative planning can be seen, in general, in China's Five Year Plans; and, in the case of renewables, in documents such as the National Medium and Long-Term Development Plan for Renewable Energy in China. Also, the corporatization of state owned enterprises (SOEs) was not simply a gradual move toward privatization, but an attempt to create national-scale holding companies where "state ownership" was in a controlling position, to develop large-scale enterprises across territorial and product sector lines, introduce advanced technology, create new products, and work toward achieving international competitiveness. In other words, the most recent advancements in industrial reforms demonstrate mixed elements of industrial policy (an effort to nurture certain industrial sectors) and public ownership as means to achieve developmental goals.

3.2.2. Gradualism and fragmentation

Gradualism is, simply put, an approach to policy-making in which changes are introduced incrementally, experimentally, and stemming from the pre-existing policy set-up (i.e., in an evolutionary fashion). In referring to the overall Chinese economic reforms, gradualism has been widely documented.

Indeed, many of the aforementioned divergences from "best practice" in the promotion of renewable electricity can be explained in light of gradualism, such as for instance: increasing but still insufficient taxation of coal; the focus on promoting installed capacity before focusing on efficiency as the goal of either mandated market policies or financial incentives; the increasing but still insufficient remuneration and duration of mandated market policies (whether tenders, independent projects, or even FITs); and increasing but incomplete regulation and enforcement of PPAs. Also, institution building is clearly underway, and the following institutional barriers could be seen as the result of gradualism: general legal insecurity; fragmentation of the bureaucracy; targets that remain non-binding; insufficient regulatory details in the Renewable Energy Law and its provisions; increasing albeit insufficient wholesale competition, or the preeminence of public ownership in generation (in spite of many steps toward liberalization of the electric sector, as well as the privatization in many other sectors). Finally, experimentation can be seen in the wide range of policies implemented: China uses (or has used) most of the policies in the toolbox, also experimenting with institutions - for instance, frequent modifications of incentives to foreign participation in generation or manufacturing.

Gradualism implies not only that China crosses the river by groping for stones, but that it might be unclear what is on the other side (what the regulatory goals are). If so, the fit of Chinese renewable energy policies and institutions with "best practice" might remain forever partial. Also, because other institutional forces, beyond the transition from a planned economy to a market economy, shape Chinese policies and institutions for renewables, the divergence is considered an optimal framework.

What complicates with gradualism is the fragmented governance of energy policy in China. Though China is good at planning, energy policy and energy planning suffers from: a) lack of consistency in the overall energy planning, in other words, the sectoral planning in the overall planning is not the result of top-down and bottom-up balancing but rather the subjective expression of the responsible government agencies or their branches; b) conflicts among policies formulated by different government agencies; c) hasty and frequent changes in the planning and policy details.

3.2.3. Planned economy

China is in the transition from a planned economy to a Socialist Market Economy. Historically, together with developmentalism, Chinese governments at all levels directly participate in economic activities. Production activities are organized based on the detailed planning by the governments, but based on the demand of customers. Though China has lifted the ban on most commodity markets, market reform in energy sector is only partly conducted and incomplete at all. Actually, due to its unique role, energy can be seen as a touchstone to judge the determination of Chinese Government to embrace market system. On one hand, energy is the basic input of production and necessity of daily life so historically the government has endowed it with developmental and social equity functions. Therefore, price distortion and subsidy, especially on coal due to its dominant position in China's primary energy mix, has deep-rooted origin in China. On the other hand, most energy product and service, including oil, power supply and grid service, is deemed as "strategic importance" or "natural monopoly". Therefore to strengthen the commanding height of the state-owned economy, state-ownership and ban on private and foreign investors, is the explicit policy orientation of Chinese government. In turn, energy market reform is proven to be extremely difficult in China. By far, the price of coal for other purpose is already determined by market competition, but the supply of coal for power generation is still under the close management of the government. For wholesale price of electricity generated from coal combustion, though a co-movement mechanism is in place since 2007 to transfer the price signal to the final customers, the mechanism is at most a semi-market and power generators receive substantial subsidy, especially when the market price of coal is high. Meanwhile, the claimed goal of power sector reform, competition in wholesale power market, is not yet implemented and the wholesale power price is set based on the benchmark price of coal power, which is certainly in favor of large-scale coal power plants. In this way, the playing field for renewables is not level in China. And without a power market, the power generation and schedule plan is scheduled according to the reliability, not the marginal cost of the generation units. However, the nature of renewable generation is intermittent. In this way, even with the provision of full purchase of renewable power generation, the actual utilization of grid-connected renewables is substantially lower than they could otherwise. The planning mechanism, covering project approval, generation planning and tariff setting, is among the key obstacles hindering renewable development in China.

4. Scope of this Book

This book is prepared to provide an aggregation of knowledge on China's renewable energy industry. Except for technological aspect, industrial, economical, institutional and policy-related aspects of renewable energy will be addressed in the Chinese context. The introductory chapter will describe the overall picture of renewable energy in China, i.e., resource endowment, policy, and overall status of industrial development. Then four industries, namely wind power, solar power, biomass and ocean energy, will be discussed in separate chapters on their status quo, policy development, achievements and challenges, market outlook and investment opportunity. We choose these four industries because they are nearly commercialized (wind power and biomass power) or are promising in the coming decade (solar power), or are novel but with great potential (ocean energy). We don't include a chapter for hydropower, because it is mature; we also don't dedicate a chapter for geothermal energy because its potential is limited. Another chapter is for smarts grids and distributed energy development. Though smart grid is not renewable energy, it provides a great opportunity to make the mass deployment of renewable energy possible in China. It is also worth noticing that due to the imbalance in resource and demand, smart grids will assume several special missions in China but the current institutional arrangement will pose challenges to the future development. Considering the dilemma of China's PV industry, we devote a chapter on Golden Sun Program, a government initiative to promote domestic solar power market. Another chapter will address a special topic, rural energy poverty and renewable energy, which will discuss the social and developmental aspects of renewable energy. Finally, the last chapter will address another hot topic, transition to a renewable energy system. The methodology from social-technical systems transition will be employed to discuss the vision, strategy, management framework, and policy package to deliver the transition in China.

References

[1] National Statistical Bureau of China (NSBC). (2013). China Statistical Book 2012. Available from: http://www.stats.gov.cn/tjsj/ndsj/2012/indexch.htm.

[2] Yuan, J., Kang, J., Zhao, C. & Hu, Z. (2008). Energy consumption and economic growth: Evidence from China at both aggregated and disaggregated levels. *Energy Economics*, *30*(6), 3077-3094.

[3] Yuan, J., Kang, J., Yu, C. & Hu, Z. (2011). Energy conservation and emissions reduction in China—Progress and prospective. *Renewable and Sustainable Energy Reviews*, *15*(9), 4334-4347.

[4] IEA. (2012). *CO2 Emissions from Fuel Combustion* (2012 Edition), Paris.

[5] IEA. (2009). *World Energy Outlook 2009*. Paris: OECD Publishing.

[6] The PEW Charitable Trusts. The PEW Environment Group. *Who's winning the clean energy race?*2011 edition. 2012. Available from: http://www.pewenvironment.org/uploadedFiles/PEG/Publications/Report/FINAL_forweb_WhoIsWinningTheCleanEnergyRace-REPORT-2012.pdf.

[7] Hong, L., et al. (2013). Assessment of China's renewable energy contribution during the 12th Five Year Plan. *Energy Policy*, Available from: http://dx.doi.org/10.1016/j.enpol.2013.07.110 (in press).

[8] IEA Statistics © OECD/IEA. Energy Statistics and Balances of Non-OECD Countries, Energy Statistics of OECD Countries, and Energy Balances of OECD Countries.

[9] IEA Statistics © OECD/IEA. Energy Statistics and Balances of Non-OECD Countries, Energy Balances of OECD Countries.

[10] Chinese National Committee on Large Dams. Water resources potential and developments in China, 2010.

[11] Li, J., Gao, H., Shi, P., Shi, J., Ma, L., Qing, H., et al. (2007). China wind power report-2007. Beijing. China Environmental Science Press.

[12] CWERA. (2009). Center for wind and solar energy resources assessment. Available from: http://cwera.cma.gov.cn/Website/index.php?ChannelID=39&NewsID=1434.

[13] Zhang, P., et al. (2009). Opportunities and challenges for renewable energy policy in China. *Renewable and Sustainable Energy Reviews*, *13* (2), 439-449.

[14] Wang Q. & Chen Y. (2010). Status and outlook of China's free-carbon electricity. *Renewable and Sustainable Energy Reviews*, *14*(3), 1014-1025.

[15] Yuan, J. & Hu, Z. (2011). Low Carbon Electricity Development in China—an IRSP Perspective based on Super Smart Grid. *Renewable and Sustainable Energy Reviews*, *15*(6), 2707-2713.

[16] Liu, W., Lund, H., Mathiesen, B. V. & Zhang, X. (2011). Potential of renewable energy systems in China. *Applied Energy*, *88*(2), 518-525.

[17] National People's Congress (NPC). (2005). The Renewable Energy Law of the People's Republic of China (Original 2005 Version), available from: http://news.xinhuanet.com/energy/2008-06/18/content_8392646.htm (in Chinese) and http://www.npc.gov.cn/englishnpc/Special/CombatingClimateChange/2009-08/25/content_1515301.htm(in English).

[18] Schuman, S. & Lin, A. (2012). China's Renewable Energy Law and its impact on renewable power in China: Progress, challenges and recommendations for improving implementation. *Energy Policy*, *51*, 89-109.

[19] Zhao, X., Wang, F. & Wang M. (2012). Large-scale utilization of wind power in China: Obstacles of conflict between market and planning. *Energy Policy*, *48*, 222–232.

[20] Luo, G-L., Zhi, F. & Zhang, X. Y. (2012). Inconsistencies between China's wind power development and grid planning: an institutional perspective. *Renewable Energy*, *48*, 52-56.

[21] National People's Congress. (2009a). Amendments to the Renewable Energy Law of the People's Republic of China, available from: http://www.npc.gov.cn/huiyi/cwh/1112/2009-12/26/content_1533217.htm.

[22] Zhao, X., Zhang. S., Yang, R. & Wang, M. (2012). Constraints on the effective utilization of wind power in China: An illustration from the northeast China grid. *Renewable and Sustainable Energy Reviews*, *16*(7), 4508-4514.

[23] Yuan, J., et al. (2012). Managing power system transition in China. Reprinted from *Renewable and Sustainable Energy Review*, *16*, 5660-5677. Copyright (2013), with permission from Elsevier.

[24] García, C. (2011). Grid-connected renewable energy in China: Policies and institutions under gradualism, developmentalism, and socialism. *Energy Policy*, *39*(12), 8046-8050.

In: Renewable Energy in China
Editors: J. Yuan, G. Luo and J. Chen

ISBN: 978-1-63117-610-4

Chapter 2

THE WIND POWER SUPPLY CHAIN IN CHINA

Jiahai Yuan [*]***, Shenghui Sun and Jiakun Shen***
School of Economics and Management, North China Electric Power University,
People's Republic of China

ABSTRACT

The wind power industry has experienced swift development and gradually moved towards maturity in China. However some hiding issues have appeared and threatened its sustainable development. In this chapter, we construct an analysis framework in the perspective of supply chain and present a thorough and comprehensive review on China's wind power industry. First we identify key stakeholders along the supply chain and discuss their concerns. Then we summarize the evolution of related policies in both the upstream and downstream parts of the supply chain. Analysis on the concerns of stakeholders and the existing policies will help understand the progress and challenges of wind power in China. We find that inability of independent technique innovation, unordered and unbalanced pattern of wind farm development, de-motivation of the power grid companies, incompatible technical codes of power grid operation, and overlooked demand response are among the main issues. Policy recommendations are proposed to address these problems.

Keywords: Wind power, supply chain, China

1. INTRODUCTION

As the biggest developing country as well as the largest GHG emitter in the world, China has been perplexed by issues of too much fossil energy consumption and high environmental cost. Therefore, striving to develop non-fossil energy is an important measure for China to ensure energy security, adjust primary energy structure and respond to climate change.

[*] Email: yuanjh126@126.com.

Accordingly, China has pledged a 15% clean energy target and formulated several clean and renewable energy plans [1].

Wind power is among the most promising green energy options due to its superior features of abundance, renewability, wide distribution and zero pollution. In recent years, wind power has been growing rapidly worldwide and plays an essential role in the diversification of the generation mix.

In the end of 2011, the cumulative installation of wind power in China reached 62.4 GW, accounting for one quarter of world total and holding the top position in the world (Fig. 2-1). In 2011, wind power generation could satisfy electricity demand of about 47 million households in China, representing energy conservation of 22 million tons standard coal equivalents (sce) and CO_2 abatement of 70 million tons [2].

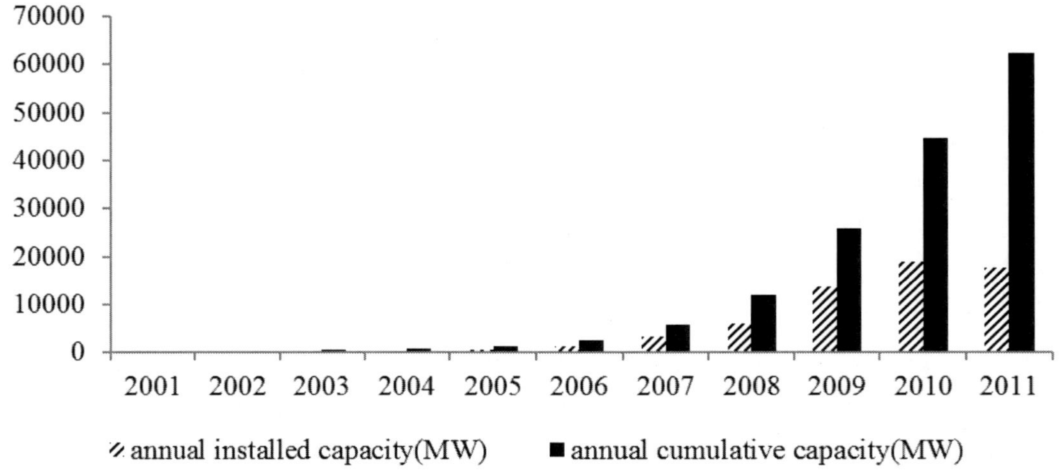

Figure 2-1. Annual installed and cumulative wind power capacity in China, 2001-2011.

However, after several years of rapid development, some issues have gradually appeared and called for a thorough and comprehensive review on the industry. The declining growth of wind turbine manufacturers and their deteriorated economic performance, the exacerbated puzzle of grid integration, and the unprecedented high rate of wind power curtailment etc. are among the primary issues. Since 1996 when State Development Planning Commission (SDPC) formulated the *Riding the Wind Program* to initiate wind power market in China, a large body of studies has been conducted. The existing literature can be classified into three streams. The first stream concerns with the status and issues of wind industry in different periods. Literature [3] pointed out that small market scale is the main reason of sluggish growth of wind power in China during the 1990's. Literature [4] pointed out that main issues during the early 21st century (2000-2005) are insufficient resource survey, low level of technology and equipment, high cost of power generation and inadequate supporting policy. Their comments were responded by the study in [5]. In [6] it was argued that the power grid is the bottleneck of wind power development. The planning of power grid considering wind integration and standardization of the planning scheme have also been discussed [7]. The literature [8] identified several main obstacles affecting the sustainability of the industry, including regulatory factors, grid integration, and technological innovation. Zhang and Li [9] identified four challenges, including uncoordinated development between the wind farms and power grids, lack of technical codes for wind power integration, unclear role on the grid companies' responsibility for grid connection, and inadequate economic incentives for grid companies. Regarding the low capacity factor of wind power, discordance between wind farm development and grid planning is widely discussed in literature [10-12]. In [13] Zhao et al. identified 19 factors concerning the strengths, weaknesses, opportunities and threats associated with China's wind power industry.

The second stream concerns with the development pattern of wind power industry. In [14] it was claimed that development of large-scale wind farms in resource enrichment regions should be the priority in China. Literature [15] proposed a pathway for developing domestic wind power equipment manufacturing. Literature [16] analyzed the dynamic mechanism of wind supply chain and proposed an operation model. Literature [17] addressed three issues, including pricing policies, transmission capacity and the structure of the equipment manufacturing industry and argued that state dominance partly contributed to these issues. Literature [18] discussed the imbalanced regional distribution of wind farms and its pressure on grid-access. Literature [19] proposed a performance framework to evaluate China's wind power policy and addressed the issue of power tariff reform.

The third stream concerns with the policy and function division between government and market [20]. Literature [21] formulated an improved dynamic Diamond Model to analyze the competitiveness of China's wind power industry and concluded that it is the duty of the government to strengthen the competitiveness of wind power against coal-fired power. Literature [22] presented a comprehensive review on policy progress during the 11th Five-Year-Plan (FYP) period and identified three defects in the policy design. Literature [23] explained how political and institutional factors have determined the relative successes and failures of China's wind power policy over the period 2005-2011. Their finding is that the progress can be attributed to the political motives and institutional arrangements of the Chinese government as well as the institutional changes. But they also argued that improper institutional arrangements lead to the low proportion of grid integration and the rising trend of wind turbine incidents.

The existing literature has offered valuable inspiration to the formulation of wind power policy in China. However, most of the existing literature is conducted in the perspective of macro policy, while the internal structure of wind power industry has seldom been addressed. Though issues relevant to the downstream part (wind farm and the grid-access) have been extensively discussed, few efforts have been devoted to the upstream part and the interactions between them. Literature [24] proposed to assess renewable energy development in a perspective of supply chain. As a matter of fact, current literature on China's wind power industry has clearly indicated that any links in the supply chain have direct impact on the industry. In this sense, study on the industry from a perspective of supply chain is a new and promising direction and can shed new insight for the industry.

The aim of this chapter is to conduct such an analysis, with an emphasis on the stakeholders along the supply chain. The remainder of the chapter is as follows. Section 2 will introduce the basic structure of wind turbine and outline the methodology of this chapter. Section 3 will discuss the stakeholders along the supply chain and their concerns. Section 4 will review the related policies. Section 5 will analyze the status and issues of the supply chain. Sections 6 will propose policy recommendations and Section 7 concludes.

2. METHODOLOGY

2.1. Wind Power Generation and the Wind Turbine System

For the purpose of our study, it is necessary to briefly introduce the wind turbine system and the process of energy utilization from wind power generation.

A wind turbine includes a cabin, a set of blades, a tower, a foundation and others. In the cabin there are generator and high and low speed shafts connected by gearbox to convert the rotor motivity driven by the wind into electricity. Below the cabin is a yaw system to ensure that the cabin is windward to utilize wind energy at the maximum efficiency. The whole cabin is erected on the foundation by a tower to capture more wind power at higher position. There are several other important systems to improve generation efficiency and guarantee safe operation, including a braking system, a cooling system, a hydraulic system and a control system, etc. (Fig. 2-2).

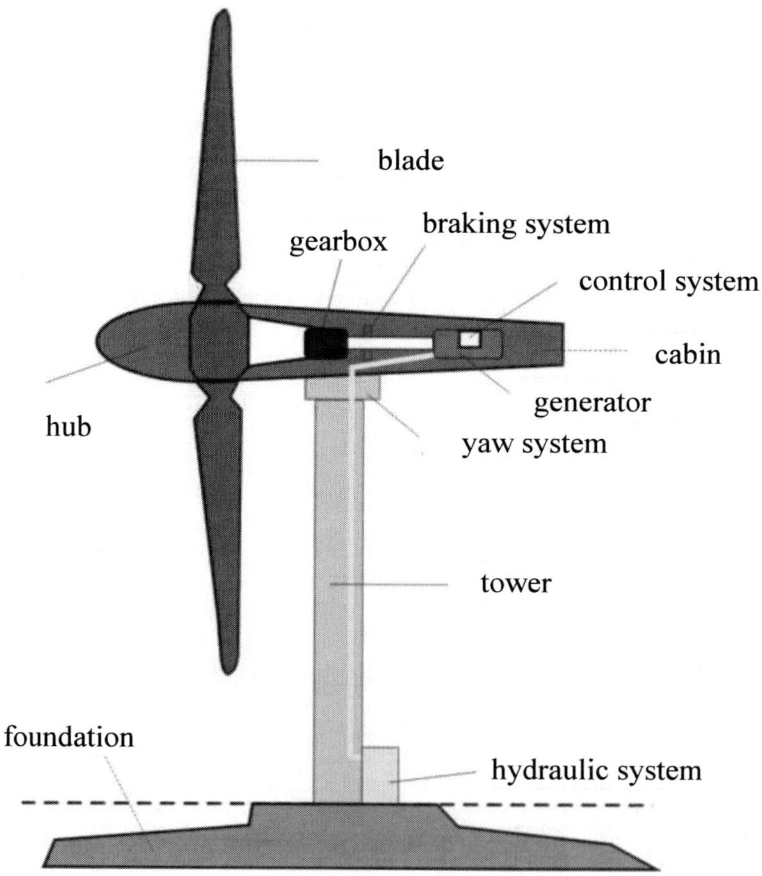

Figure 2-2. Wind turbine system.

In China, the manufacturing and supply systems for the wind turbine and its key components and parts are gradually formed. In the manufacturing system for the key components, manufacturing capacity of the blade, gear box and generator is improving quickly and a whole supporting system is forming. But some key components still cannot be manufactured independently in China. For example, the principal bearings for the 2-MW wind turbine are dependent on import. Some other, such as electrical control and hydraulic systems are also dependent on import. Control system can be domestically manufactured partly but it is still the bottleneck of wind turbine manufacturing in China.

The basic principle of wind power generation is to convert the captured wind power into mechanical power, which in turn drives the rotor of the generator to cut magnetic lines and generates electricity for final usage (Fig. 2-3). The rotor blades of the generator are rolled with the adjustment of the yaw motor and drive the low-speed shaft. Then high-speed shaft is driven by the low-speed shaft via the gearbox and rotates the generator. The generated electricity is then transformed via the transformer and power electronics equipment. Finally electricity at proper voltage is transmitted to the customers by the transmission and distribution system. The control system, which is responsible for measuring of wind speed and direction, regulating the pitch between the rotor blade and generator, and adjusting the rotary speed and the high-speed shaft, plays a critical role in the process. Most of the electric power will be consumed by the industrial, commercial or residential customers. However, when there is no much demand some can be stored, by hydrogen, cell fuel, and pump hydropower or storage boiler. On the other side, when there is excessive demand and the supply cannot meet up with it, demand response can be used to balance them.

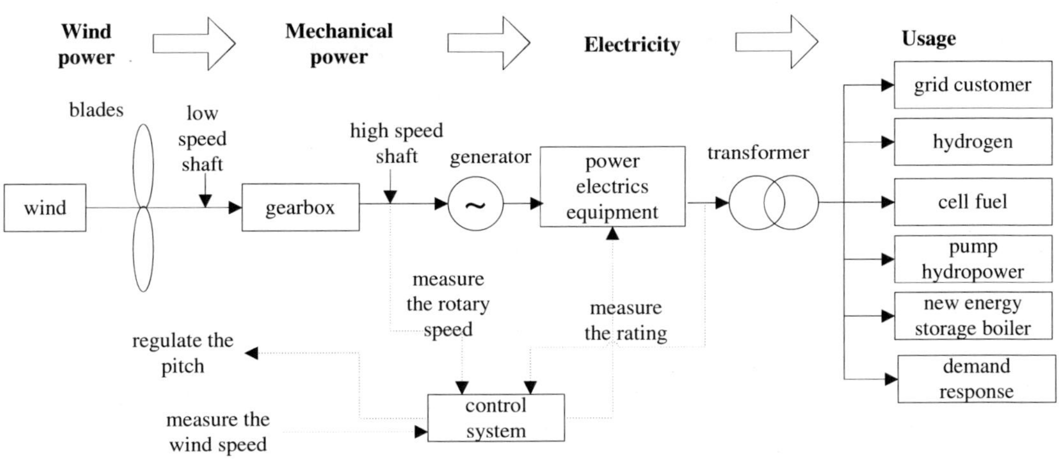

Figure 2-3. Wind power generation and utilization.

2.2. The Framework of the Analysis

The wind power supply chain includes raw material suppliers, components manufacturers, wind turbine manufacturers, wind farm developers and/or operators, grid operators and the related service suppliers. (Fig. 2-4)

In a perspective of supply chain, the wind power industry can be divided into two parts, the upstream and the downstream. The upstream is consisted of the raw materials suppliers, the components and parts manufacturers, the technology servicers, the wind turbine manufacturers and the wind farm developers. The raw materials suppliers provide raw materials to the components manufacturers, then the components manufacturers and technology servicers provide components or services to the wind turbine manufacturers. Then with the turbine supply and the engineering services, the developers conduct investment, siting and infrastructure activities to build the wind farms. The downstream is consisted of the wind farm operators, the grid companies, the customers and future customers/distributors equipped with energy-storage facilities. The grid companies provide grid-access service to the wind farms and invest in the transmission system. The electricity generated from wind power is transmitted to end-use customers via the transmission system or stored by energy storage facility.

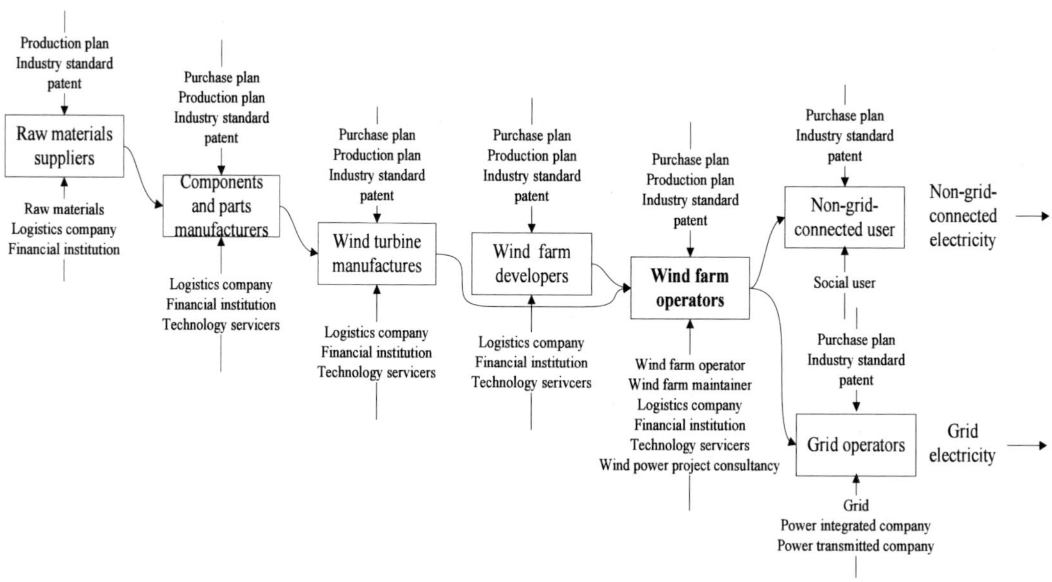

Figure 2-4. Wind power supply chain.

The nodes of the supply chain are connected with each other via logistics, information flow and cash flow. The performance of the supply chain depends upon the performance of individual nodes and the coordination among all the stakeholders as well.

3. Stakeholders and Their Concerns

In this section, we will identify the stakeholders along the wind power supply china in China. The upstream is the supply chain for wind turbines. There are seven key stakeholders, including central government, the R&D institutions, the patent owners, the components and parts manufacturers, the turbine manufacturers, the wind farm investors and local governments (Table 1).

Here central government includes not only State Development and Reform Commission (SDRC), and National Energy Administration (NEA), who are responsible for industry policy, energy planning and price policy, but also Ministry of Finance (MOF) and State Administration of Taxation (SAT), who are responsible for the formulation and implementation of fiscal and tax policies. Central government plays essential roles in the following ways. First, by formulating related laws and planning for renewable energy, strong signal for wind power development is provided. Second, by formulating and implementing supportive industry policies, domestic manufacturing capacity is fostered. Last, by favorable price and tax policies, a strong domestic wind power market is successfully created.

Table 2-1. The upstream stakeholders in China's wind power supply chain

	Stakeholders	Function	Concerns	Related with
Entire supply chain	Government	Industry development	Fostering new strategic industry	-
		Energy planning	Energy security and stable energy supply	
		Renewable energy utilization	Environmental and ecological issues	
Equipment manufacturing	Technology provider& Patent owners	R&D of new technologies Supply of key technology to manufacturers and wind farm operators	Profit sharing from technology service, protection of intellectual property	Wind turbine manufacturers
	Components and parts manufacturers	Manufacturing and supply of required components and parts for wind turbine	Market share, competiveness and profit	Other components and parts manufacturers
		Reducing product cost by technology R&D	Imitation, R&D and diffusion of key technologies	Patent owners
	Wind turbine manufacturers	Assemble wind turbines	Market demand, market share and profit	Competitors; wind farm investors; components suppliers
		Engineering service for building wind farms	Siting of wind farms, profit of the service delivered	
		Technology R&D	Production cost, acquirement of new technology	Patents owners
Wind farm development	Wind farm investors	Detailed planning of wind farms	Wind resource, land price, local government policy, grid access	Local government, wind turbine suppliers, grid companies
		Investment of wind farms	Total investment, expected profit	
	Local governments	Regional wind power resource survey and planning, policy support to wind farm development	Potential contribution to local economy, tax and employment	Central government, grid company, wind farm investors

Table 2-2. The downstream stakeholders in China's wind power supply chain

	stakeholders	Function	Concerns	Related with
Entire supply chain	Central government	Overall renewable energy policy	Share of renewable in the energy mix	-
		Specific pricing, tax and fiscal policy pertinent to wind power	Commercialization of wind power	
Wind power generation	Patent owners	Providing technology service to wind farms and power grids	Profit sharing from technology service, protection of intellectual property	Wind farms and power grids
	Wind farm operators	Wind power generation	Capacity utilization, dispatch planning	Power grid companies
		Operation & Maintenance	Cost reduction, safe and efficient operation	Patent owners
Power transmission and distribution	Power grid companies	Grid connection of wind farms	Investment requirement for the transmissions system, planning of power grid, profit (difference between retail tariff and wholesale generation tariff)	Wind farms, customers, central government
		Dispatch of the power grid	Impact of different power plants to the safe operation, response sped and price of power plants	Local governments, wind farms
		Application of Smart Grid (and the related technologies)	Benefit of Smart Grid to power grid, role of power grid	Customers, wind farms
Power consumption	Customers	Customers connected to power grid	Stability and price of electricity	Grid companies
		Isolated customers	Price of wind power and distance from the wind farm	Wind farms
Entire downstream supply chain	Local governments	Regional power generation planning, policy support to wind farm operation	Stability of regional power supply, wind farm's contribution to local economy	Central government, grid companies, wind farms

Local governments play facilitative roles in the development of wind power. Because most of wind resources are located in undeveloped areas, local governments have strong incentive to develop wind farms in their administrative regions. For them, the biggest concerns are the job opportunity and the contribution to local revenue incurred by new investment. Typically, favorable loan support, tax exemption policy and low cost land etc. are provided to attract wind power investors. But local governments also have their roles in the unordered development of wind power. For example, under the existing regulation system, wind power projects with capacity above 50 MW will be checked and approved by SDRC but projects with capacity less than 50 MW will be approved by the local governments. To circumvent the strict and lengthy approval procedure of the central government, the developers usually decompose large projects into several 50 MW projects and then get them approved by the local governments. Naturally, these projects are not listed in the national renewable energy planning. As a result, grid connection will be a puzzle. It is also worth noticing that provincial governments have a saying in the arrangement of annual power generation planning and they favor secured power supply, instead of wind power generation, which is intermittent [12].

The downstream is the supply chain of wind power and the key links are the wind farms (power generation), the grid companies (wind power integration, transmission & distribution) and the customers (utilization). There are eight stakeholders, including the R&D institutions, the patent owners, the wind farm operators, the grid companies, the customers connected to public grids, the isolated customers, the local governments and central government (table 2-2).

Grid companies play a vital role in the downstream supply chain. According to the provision of *Renewable Energy Law*, grid companies are obliged to fully purchase wind power generation under suitable grid access conditions. But it doesn't necessarily imply that grid companies have appropriate incentive to provide the grid access and transmission services for wind farms. The biggest concern for grid companies is the safe operation of the power grids. Wind farms, however, add uncertainty to the grid operation. In the operation of the power systems, controllable power supply dominates the generation mix. Typically, nuclear power and hydropower serve for base load, coal power for medium load and gas power for peak load. Because of the intermittence and the difficulty of accurate output forecast, it is impossible to take wind power as a priority in the dispatching plan. Therefore, proper handling of wind power in dispatch and operation is an urgent requirement. Smart Grid is an imperative solution for it. Another concern of grid companies is the extra cost incurred by wind power. Extra transmission facility and extra reserve capacity for wind power bring forth extra cost to the grid companies. But there is no specific provision in the *Renewable Energy Law* on this issue.

Typically power customers care about the stability of power supply and its reasonable price, but rarely care about where it is from. In some developed countries like USA, mechanism as green certificate has successfully involved customers in. In many European countries like Germany and the Netherlands, a private investor can erect a wind turbine and arrange power supply directly with final customers and the grid companies are obliged to provide transmission service. But in China, the grid companies' monopoly role has essentially eliminated these possibilities.

4. Policy Progress

In this section we will discuss the relevant policies in both the upstream and downstream parts. To save space, we will simplify the specific policies in tables and only address the key points (table 2-3 and 2-4). It is worth noticing that the focus of the upstream policies is to encourage the domestic manufacturing of wind turbine while the focus of the downstream is to encourage investment of large-scale wind farms by providing strong price signal.

Table 2-3. The upstream policies for wind power in China

Time	Policy	Key points and comments
2005	Notice on relevant requirements for the management of wind farm development	✓ The key points for the approval of wind farm development are the scale of the wind farm, siting condition and the domestic manufacturing ratio of the wind turbine (no less than 70%). ✓ Special attention is attached to domestic manufacturing ratio of the wind turbine in the development of wind farm.
2007	Notice on the relevant import duty policy to implement the decision to revitalize equipment manufacturing industry by the State Council	✓ Provide exemption of import tariff and related VAT to domestic manufacturers for the import of key components and material of wind turbine. ✓ Import substitution is taken as the priority to foster domestic wind turbine manufacturing industry.
2008	Interim measure on the special fund for promoting domestic wind turbine manufacturing industry	✓ Provide subsidy of 600 RMB/KW for the first 50 units of MW scale wind turbine for qualified manufacturers. The subsidy will equally be shared between wind turbine manufacturers and key components (especially converter and bearing) manufacturers. The subsidy can only be used for the R&D of new wind turbine. ✓ Public money is used to provide incentive for product innovation.
2009	Decision on restraining over capacity and guiding the healthy development of some industries	✓ Strictly control the blind expansion of wind turbine manufacturing. Forbid the provisions in the bidding specification that the wind turbine must be manufactured locally. Establish and perfect the standard of wind power equipments and the related test and certification system. Enhance the research on the technology roadmap of wind power and technology R&D of offshore wind power. Perfect the quality control system for the wind power manufacturing. ✓ Guide the healthy development of domestic wind turbine manufacturing after rapid market expansion.
2010	Exposure draft on the admittance standard for wind power manufacturing industry	✓ Formulate six conditions for establishing wind turbine manufacturing enterprise: no less than 30% of self-fund in the whole investment; equipped with all the facility to produce 1 GW/year at no less than 2.5-MW unit for the new investment, installation performance of no less than 0.5 GW for the expansion investment; five years of business experience in large-scale mechanical and electronic industry for the new enterprise; transportation convenience; full supply chain matched with wind turbine manufacturers. ✓ Increase the market concentration and improve the competitiveness of domestic manufacturers.
2011	Notice on strengthening safety management of wind farm and containing accidents of wind farm operation	✓ Formulate retrofitting plan to the wind turbine units that cannot satisfy the condition of low voltage ride through (LVRT). ✓ The government begins to pay attention to the impact on power grid when the scale of grid accessed wind power accounts for significant portion of total generation capacity in some regional power grids.

Table 2-4. The downstream policies for wind power in China

Time	Policy	Key points and comments
2005	Renewable Energy Law	✓ Require that grid companies fully purchase the renewable generation and provide grid-access service to renewable generators. The surcharge of purchased price of the grid company should be allocated by all customers in retail tariff. ✓ Provide legislative base for renewable energy development and strong incentive for wind power.
2006	Trial measures on the generation price and cost allocation of renewable energy	✓ Implement government referential price for electricity generated from wind power and the price will be determined by department in charge by tender offering. An additional fee will be collected from electricity customers to solve the issue of high price of renewable power generation. ✓ Provide the practical way to implement provisions of the Renewable Energy Law
2007	Supervision measures to power grid companies on the full purchase of renewable power generation	✓ Grid companies are liable to fully purchase the grid-accessed renewable power generation in its coverage area and renewable generators should provide the necessary aid. Grid companies are responsible for disclosure of the related information. ✓ Provide practical measure for gird-access of wind power.
2008	Notice on the integrated utilization of resource and VAT policy of other products	✓ Wind power operators can enjoy the policy of 50% exemption of VAT. ✓ Provide fiscal support to wind farm.
2008	Notice on the preferential income tax catalog for enterprises in the field of public infrastructure	✓ Allow wind farm operators to deduce the input VAT of the newly-installed turbine at the current VAT rate. The wind farm operators are eligible for full exemption of income tax for the first three years of regular business operation and half exemption for the second three years. ✓ Provide fiscal support to wind farm.
2009	Notice on policy to perfect wind power generation price	✓ Divide the whole nation into four resource areas and set the benchmark price of 0.51, 0.54, 0.58 and 0.61 RMB/KWh. Offshore wind price will range between 0.62 and 0.97 RMB/KWh and is determined by the combination of bidding and approval. ✓ Simplify the price mechanism of onshore wind power and set up the pricing mechanism for offshore wind power.
2011	Notice on the approval of first batch of wind power projects during the 12th FYP period	✓ Approve wind power projects of 28.83 GW during the 12th FYP period and provide a detailed project list. The unlisted projects cannot be approved and accessed to power grid and cannot enjoy the renewable subsidy. ✓ Formulate planning of wind power during the 12th FYP period.

5. Status and Issues

Overall, there are lots of imbalances in the supply chain. In the upstream supply capacity is the policy priority but R&D capability is weak. In the downstream the speed of wind farm development is fastest in the world but grid integration and final usage are poor. In one word, the mismatches along the supply chain pose serious threats to the commercialization of wind power in China.

In the upstream, the manufacturing of MW-unit wind turbines is still in the initial stage. By means of conducting joint R&D with foreign manufacturers or gaining license from them, domestic manufacturers are capable of producing wind turbines at MW scale. But the technology digestion is weak and mass production is still problematic. Meanwhile, the capacity of domestic manufacturing of key components, especially the blade, gear box and main bearing is weak and cannot meet the requirement of domestic market. In the downstream, the development of wind farms is rapid. Investors, especially the state-owned (SOE) national generators are keen on wind power development and bring strong demand for wind turbines. But the final usage of wind power generation is restrained by the grid connection and other institutional issues.

5.1. The Upstream Supply Chain

The upstream part of wind supply chain can be divided into components and turbine manufacturing (Fig. 2-5).

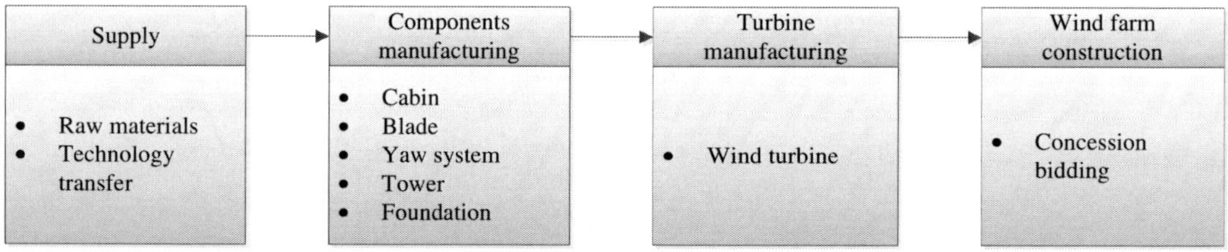

Figure 2-5. Upstream wind power supply chain.

5.1.1. Wind Turbine Manufacturers

5.1.1.1. Status

In 2011, there were 29 wind turbine suppliers in China. The number was 38 in 2010 and 43 in 2009. It is worth noting that "supplier" differs with "manufacturer" here and means company who sells wind turbine. There are about 70 registered manufacturers in China though not all of them contribute to domestic supply every year.

Wind turbine manufacturing is an industry with obvious economy of scale. Only those manufacturers with solid capital base can attract and aggregate talents and decrease the production cost by technology innovation. So it is not surprising that first movers possess scale and technology advantages and are always the successful bidders in the wind power projects. Therefore, turbine manufacturing is an industry with high market concentration. In newly increased installation of 2011, five top suppliers were Goldwind 3600MW (20.4%),

Sinovel 2939MW (16.7%), United Power 2847MW (16.1%), Mingyang 1177.5MW (6.7%) and DTC 946MW (5.4%). While in the end of 2011, top five suppliers of cumulative installation were Sinovel 12977MW (20.8%), Goldwind 12678.9MW (20.3%), DTC 6898MW (11.1%), United Power 5282MW (8.5%) and Vestas 3565.5MW (5.7%) (Table 2-5)[2]. Top five manufactures hold more than 60% of the total market share in China (refer to the Appendix for the company names appeared in the text).

Table 2-5. Top 20 wind turbine suppliers in China in 2011

Unit: MW	Newly increased supply in 2010	Cumulative supply in 2010	Newly increased supply in 2011	Cumulative supply in 2011
Goldwind	3735	9078.85	3600	12678.9
Sinovel	4386	10038	2939	12977
United Power	1643	2435	2847	5282
Mingyang	1050	1945.5	1177.5	3123
DTC	2623.5	5952	946	6898
XEMC	507	1089	712.5	1801.5
Shanghai Electric	597.85	1073.35	708.1	1781.5
Vestas	892.1	2903.6	661.9	3565.5
CCWE	486	682.5	625.5	1308
CSR Zhuzhou	334.95	465.3	451.2	916.5
GE	210	1167	408.5	1575.5
Haizhuang	383.15	479.25	396	875.3
Windey	129	723	375	1098
Gamesa	595.55	2424.3	361.6	2785.9
Envision Energy	250.5	400.5	348	748.5
Yinxing Energy	154	252	221	473
SANY	106	143.5	179.5	323
XJ Wind power	22	26	166	192
HY	161.64	295.08	151	446.1
Suzlon	199.85	805.1	96.2	901.3
Others	460.9	2354.47	259.3	2613.7
Total	18928	44733.3	17630.8	62364.2

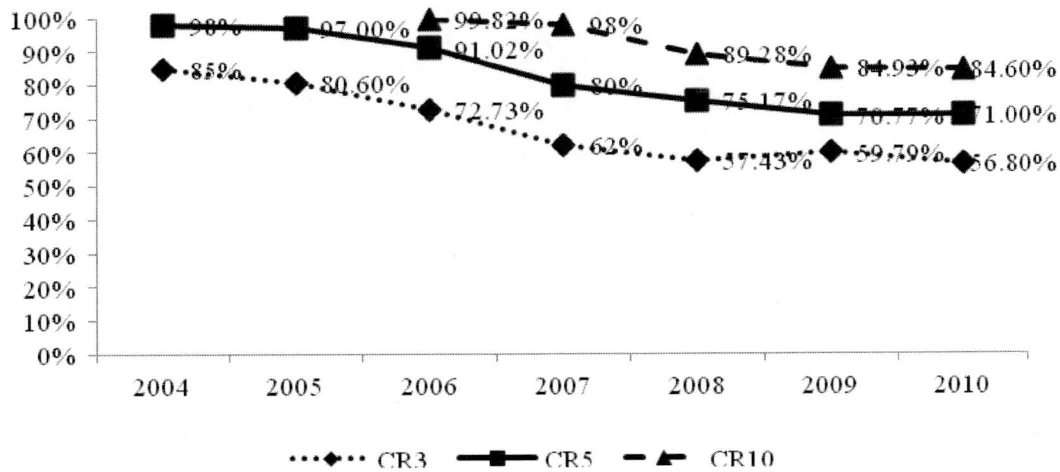

Figure 2-6. Market concentration of wind turbine manufacturing in China.

Due to rapid market expansion, more companies enter into the sector of wind turbine manufacturing. The market concentration, at the same time, declines slightly year by year (Fig. 2-6) [26].

5.1.1.2. Issues

The first issue is the lack of national technology standard. Though manufacturers can preliminarily grasp the know-how for producing MW-scale turbines, the generality and expansibility remains problematic due to the lack of national standard.

The second issue is overcapacity. In 2007 there were only about 30 wind turbine manufacturers in China. However in just one year more than 40 companies entered into the industry after Goldwind, the leading domestic manufacturer who went public in the end of 2007. The blind entrance results in overcapacity. Now top 10 suppliers hold about 90% of the market share while the rest of 90 manufacturers compete for the remaining 10% and most of them can never gain their orders.

The last one is quality issue. Most of the wind turbine technology is imported and has not gone through a scientific appraisal and re-innovation process to adapt to local conditions. Additionally, because supply falls short of demand for several years, many wind turbines are put into volume production even without trial operation. When these turbines are connected to the power grids, they certainly will pose potential quality issues and safety hazards.

5.1.2. Components and Parts Manufacturers

5.1.2.1. Status

Table 2-6. The trend of wind power technology development

Field	Development trend
Wind turbine	✓ 1.5 MW and above unit is the mainstream model in the market ✓ High-power offshore wind turbine is leading edge of the market ✓ 10 MW offshore wind turbine has been successfully developed
variable-speed operation	✓ Possess properties of high power generation efficiency, flexibility to variable wind speed and low production cost ✓ German company Enercon takes the leading position
generator	✓ The mainstream technology currently is double-fed, pitch and variable speed generator ✓ Stalled, variable-pitch and variable-speed generator is not the mainstream technology, but it is mature and has stable performance and much O&M experience ✓ Direct-driven, permanent magnet, variable-pitch and variable-speed generator is the future direction
Gearbox	✓ Direct-driven without gear box can efficiently improve the system efficiency and operation reliability
Blade	✓ Development from fixed to variable pitch is the direction ✓ Detachable cluster is the direction adapted to the trend of enlarged blade ✓ Raw material of the blade will developed from current resin reinforced by glass fiber to high intensity and light weight carbon fiber
Electronic control device	✓ Application of advanced control theory ✓ Non-impact speedy grid connection ✓ Remote monitor and control ✓ Independent blade control ✓ Isolated wind turbine or weak grid operation ✓ Hybrid control of wind power and PV

The components and parts manufacturing in China is growing quickly and the production and supply system is improved gradually. Now in China a system capable of manufacturing key components as the blade, gear box, generator, variable yaw system, wheel hub and tower is established. However, domestic components manufacturing can hardly keep up with the trend of global technology development. Table 2-6 provides a brief of the trend.

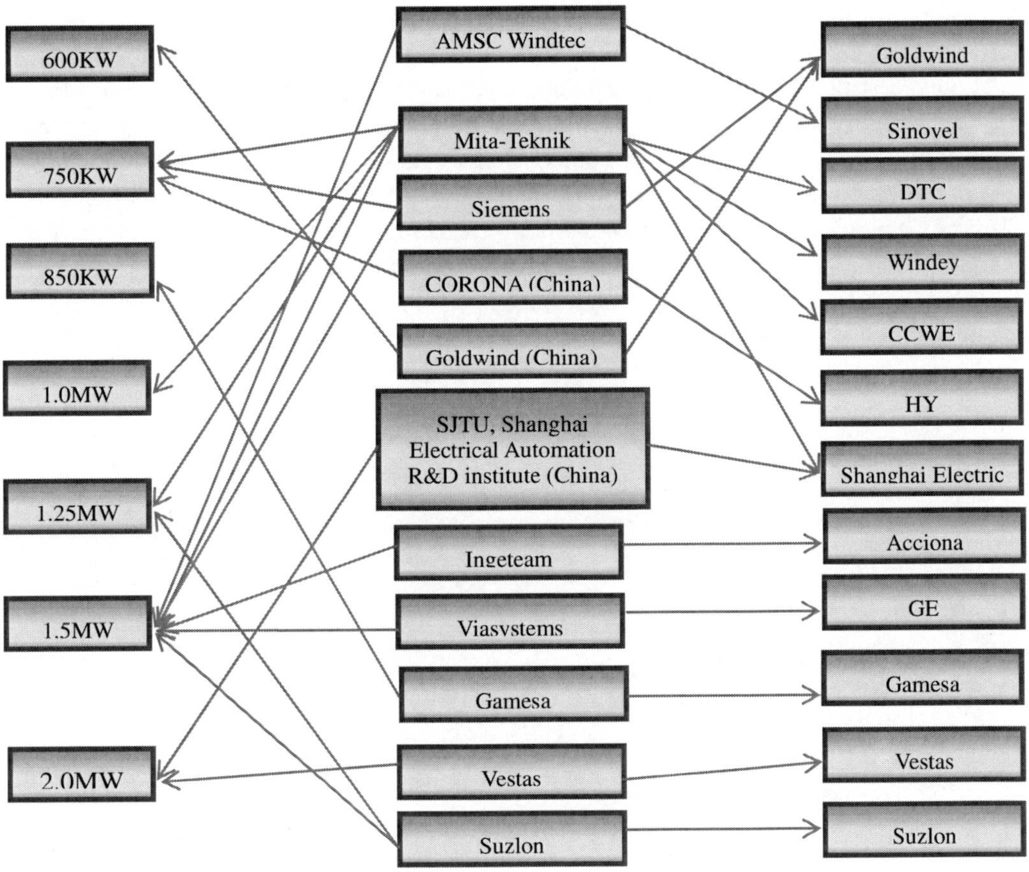

Figure 2-7. Mating among electronic control and wind turbine companies.

Since 2008, the supply of components and parts can meet the market demand in China. Some of them are supplied by domestic manufacturers while others are supplied by factories established by foreign companies. Among them, more than twenty companies manufacture blades or conduct R&D and trial production activities for the blades. The production capacity for blades in China is growing quickly and an integrated supporting system is formed. Representative companies like HT, Zhongfu Lianzhong, and SHFRP etc., have realized the volume production of high-power wind turbine blades by means of technology import or joint-design. For gear boxes, more companies have entered in but the production capacity grows very slowly. Most of the gear box manufacturers in China have years of experiences in the manufacturing of large-scale gear boxes, for example NGC, CN GPOWER, and ADVANCE. The first two companies account for 80%-90% of domestic market share. For generators, the industrial foundation is sound and there are many companies operating in

China. The suppliers that provide generators for large-scale wind turbine are YONGJI, LEC, XEMC, TMT, SEC, Tianyuan Electric, DEC Dongfeng and Nanyang Electric etc.

It is worth noting that previously the main-shaft bearings, the gear box bearings and the converters were completely dependent on import but now domestic suppliers can realize volume production for them. Therefore, the restrictions on these key components are mitigated and the manufacturing cost of wind turbine as a whole is cut down. But the electronic control system is the weakest link and still fully depends on import. The main foreign suppliers are Mita-Teknik and AMSC's Windtec. Several Chinese companies are in the R&D and test production stage, for example, CORONA, SUNGROW, XJ Wind Power, NARI, Tianyuan Electric, etc. Among them, CORONA successfully developed a prototype of 1.5 MW double-fed, variable-speed and constant-frequency control system in 2006, which is the first control system with proprietary intellectual property rights and in practical operation. Fig. 2-7 provides a diagram for the mating among electronic control and turbine companies. It is clear that domestic manufacturers can only provide the electronic control systems for 600KW and 750KW turbines, but MW-unit turbines and above are still dominated by foreign companies or their subsidies in China [26].

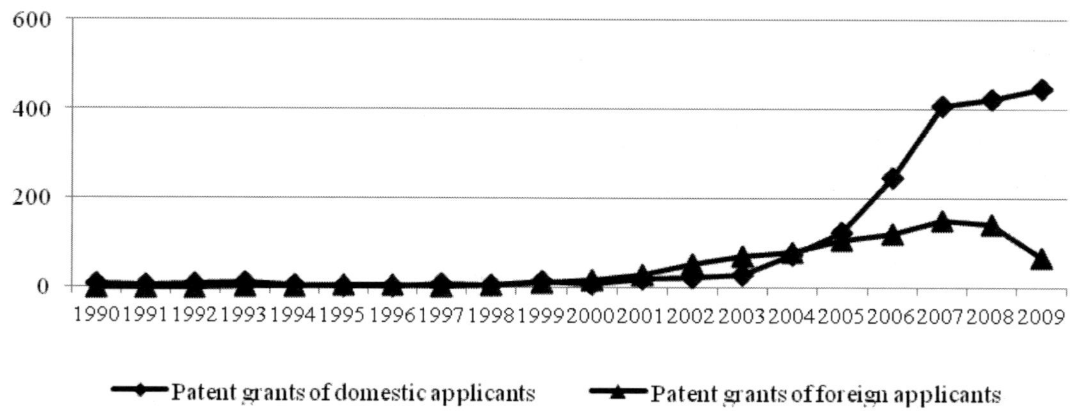

Figure 2-8. Patent application in wind turbine industry.

Analysis on patent application can provide interesting evidence. Fig. 2-8 compares the patent applications for wind turbines between domestic and foreign applicants. It is evident that the quantity of applications has grown rapidly since 2001 and foreign application contributed most of the growth before 2005, while after 2005 domestic applications began to exceed foreign applications [25]. But it doesn't necessarily imply that the innovation capability of Chinese wind turbine manufacturers catches up with their foreign counterparts since 2005. Fig. 2-9 and 2-10 present the patent applications from the strongest domestic and foreign applicants [27]. It is clear that patent applications from foreign applicants have absolute advantage in not only the quantity requested, but also the continuity of development. While for domestic applicants, their applications are active only in discrete year and lack continuity. The type of applicants is also revealing. For the Chinese applicants, individuals account for 50% of total applications, universities and research institutes account for 20% while companies account for the rest of 30%. Hence the innovation capability of Chinese companies is weak. But for foreign applicants, companies account for 81.6% of the total applications, which is strong evidence that foreign companies are the driving force of innovation in the turbine manufacturing.

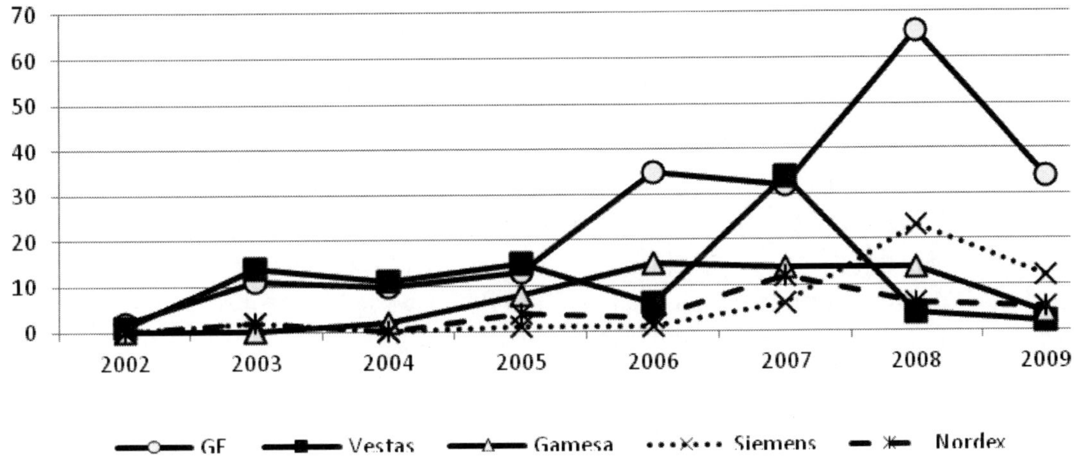

Figure 2-9. Patent application of foreign applicants in China.

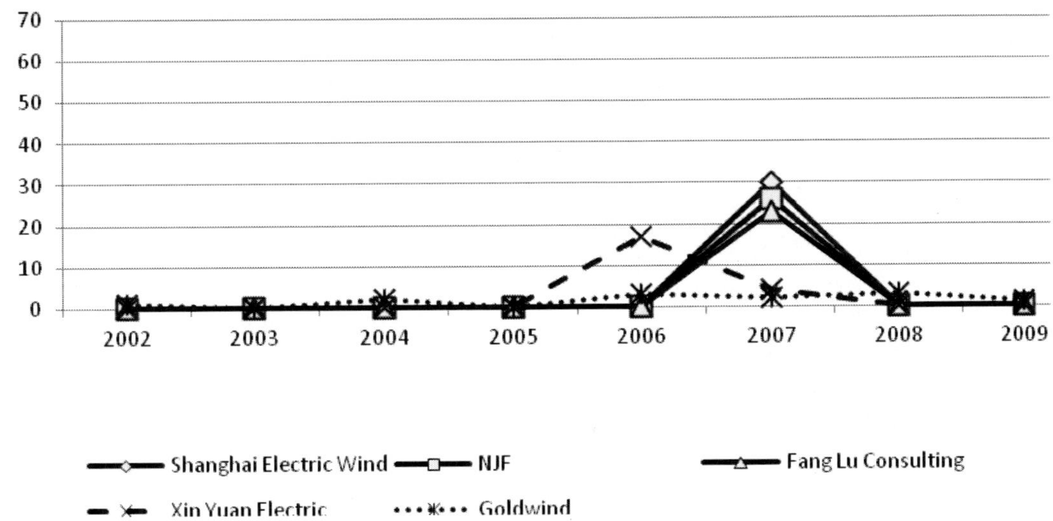

Figure 2-10. Patent application of domestic applicants in China.

5.1.2.2. Issues

Firstly, some key components are still the constraints for the release of domestic manufacturing capacity. Before 2006 the mainstream model of wind turbines in China was low-power set (600 KW or 750 KW), because it is relatively easy to produce the components. When the unit size of wind turbines increases (from KW-level to MW-level), the difficulty to master the know-how of producing key components has increased remarkably.

Secondly, the supply of some components, for instance blades and towers, is in overcapacity due to low technical threshold. It is estimated that top 10 manufacturers can fully meet the domestic market demand for blades but there are more than 50 manufacturers in China.

Thirdly, most of the components manufacturers simply rely on price competition to expand their market share, which results in vicious competition and harms the whole industry. Finally poor standardization is still a deterrent to the healthy development of the industry.

5.2. The Downstream Supply Chain

The downstream part of wind power supply chain includes wind farm, wind power generation and final utilization (Fig. 2-11). It is worth noticing that the development pattern of wind power differs greatly in China from that in Europe and the USA. Wind farms in Europe and the USA distribute widely and are mainly connected in the distribution networks and balanced with local demand. But in China, the scale of wind farms grows even bigger and most of the electricity generated from wind is consumed across regions.

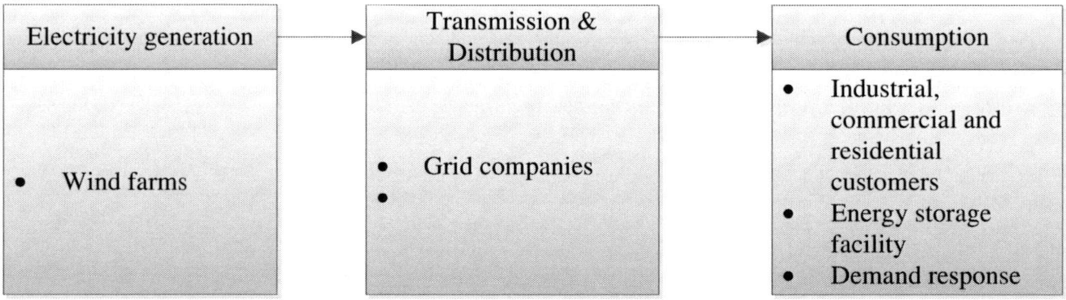

Figure 2-11. Downstream supply chain of wind power.

5.2.1. Wind Farms

Large SOE generation companies have always held the lead in wind farm development in China. Now there are more than 60 companies participating in wind power generation. Leaving aside several SOE generation companies, the rest of them are much smaller. Generally speaking, wind farm developers in China can be divided into the follows: large-scale SOEs affiliated to the central government, SOEs affiliated to the local governments, private companies, foreign-owned companies and wind turbine manufacturers involving in power generation.

In 2011, the top 10 wind farm developers installed capacity of 13.43 GW, accounting for 76.2% of annual newly-added installation. Table 2-7 provides detail information about the top 10 wind farm developers in 2011.

According to field statistics, in the end of 2011, SOE wind developers had 37.98 GW of wind power capacity, accounting for 79.4% of the total grid-connected capacity (47.83 GW); private developers had 2.18 GW, accounting for 4.6%; foreign-owned developers had 600 MW, accounting for 1.31%; the rest of 14.7% (7.05 GW) was hold by joint venture companies (Table 2-8). Among which, the top five SOE generation companies had 27.1 GW, accounting for 57% of cumulative grid-connected capacity. China Guodian Group ranked top with cumulative capacity of 9.81 GW, China Huaneng Group and Datang Group ranked second and third with 6.58 GW and 5.74 GW, respectively.

Table 2-7. Top 10 wind farm developers in China in 2011

Developers	Annual newly added capacity (MW)	Market share (%)
China Guodian Group	3860.5	21.9
China Datang Group	2235.1	12.7
China Huaneng Group	2229.0	12.6
China Huadian Group	1104.5	6.3
Guohua Group	1094.5	6.2
China Power Investment Group	866.3	4.9
China Resource Group	796.1	4.5
China Guangdong Nuclear Power	527.0	3.0
Beijing Energy Investment Holding Co. Ltd	372.0	2.1
China Suntien Green Energy	343.6	1.9
Others	4202.9	23.9
Total	17630.9	100.0

Table 2-8. Cumulative grid-connected wind capacity in China in 2011

Developers	Cumulative grid-connected capacity(MW)
China Guodian Group	9812.9
China Huaneng Group	6581.0
China Datang Group	5743.0
Guohua Group	2353.0
China Huadian Group	2837.1
China Guangdong Nuclear Power	2200.6
China Power Investment Group	2124.5
China Resource Group	1382.8
China Three Gorges Group	1340.3
Others	13460.4
Total	47835.6

5.2.1.1. Operation of Wind Farm

In this subsection, we will discuss two important issues relevant to the operation of wind farm in China, capacity utilization and CDM revenue.

The curtailment of wind power is serious in some regions and the capacity utilization factor is decreasing. The issue of grid access is gradually becoming the greatest challenge. Wind resources are mainly distributed in the Three-North regions, namely northeast, northwest and northern China. However, the demand is mainly located in eastern coastal regions. In the past few years, the development of wind power has been highly concentrated in the three-north regions. Because of the asynchrony between wind power and power grids development, the lack of adjustable backup capacity and the difficulty in trans-province electric power trade, the bottleneck of grid-access becomes acute. As a result, the curtailment of wind power increases.

Issue of curtailment became serious in 2010 when total installation reached 44.73 GW but 31% could not be connected to the power grids. In March of 2011, State Electricity Regulation Commission issued its regulation report on wind and PV power generation [28]. In the report, detailed statistical data for wind power curtailment are provided (Table 2-9). In 2010, Inner Mongolia, Jilin, Gansu and Heilongjiang provinces were among the most serious regions of the curtailment.

Table 2-9. Statistical data of wind power curtailment in China, 2010

Region	Curtailment rate (%)	Wind farms surveyed
East Inner Mongolia	22.99	98
Jilin	20.49	44
West Inner Mongolia	17.51	129
Gansu	16.99	39
Heilongjiang	14.49	59
Liaoning	10.34	74
Xinjiang	3.21	32
Hebei	3.09	74
Shandong	1.46	57
Ningxia	0.64	33
Total	11.12	639

In 2012, the National Energy Administration issued a report on annual utilization hours of wind farms in 2011 [29]. According to the report, the national average in State Grid regions was 1928 hours while it was 1801 hours in China Southern Grid regions. The region with lowest utilization was Jilin province with 1610 hours. If taking 2000 hours as the break-even point, in 2011 wind farms in only 14 provinces could earn positive profit and almost half of them were losing money. It is estimated that in 2011 the direct economic loss caused by wind power curtailment amounted to 5 billion RMB.

With the implementation of clean development mechanism (CDM) created by the Kyoto Protocol, many wind power projects in China have successfully gained support from CDM. In the end of March 2010, there were 181 wind projects successfully registered in the UN, totaling 946 MW and approximating GHG abatement of 178 million tons CO_2equivalents. Renewable energy projects can sell CERs by trade in the carbon market and acquire carbon revenues. According to World Wildlife Fund, CERs revenues roughly account for 10% of total projects investment and can increase the internal rate of return (IRR) of wind power projects from 6.81% to 9% [30]. Therefore, CERs can remarkably enhance the revenues of wind power project and increase its attractiveness. As a matter of fact, in China wind power accounted for 65% of the registered CDM projects. According to current trade price of CO_2, CDM subsidy can bring forth 0.05-0.07 RMB per KWh electricity generation. However, due to the elevated threshold of getting CDM registered, recently the percent of the rejected projects has increased substantially. And more importantly, with expiration of the first phase of the Kyoto Protocol in the end of 2012, the sudden slump of CERs price will bring great uncertainty to wind power projects in China.

5.2.1.2. Issues

According to our study, key issues on wind farm development in China can be summarized as follows.

The lack of scientific appraisal in the siting of wind farm: First, the current version of wind resource map in China is not elaborate enough to guiding the siting work. Second, some local governments hasten in wind farms projects, sometimes even without professional resource survey and appraisal. As a result, some wind farms cannot justify the initial investment due to their poor capacity factor.

The restriction of grid connection: Because of the long distance of power transmission, the extensive engineering work involved and the resulted long construction time, and more importantly, the imbalance between wind farms planning and power grids planning, some wind farms cannot be connected to the power grids and generate electricity.

The unreasonable pricing mechanism for wind power: When the prices of wind power were determined case by case by means of franchise bidding, policy of "successful bidders with the lowest bidding price" has attacked the enthusiasm of investors because it created vicious competition among them. In 2009 China introduced feed-in-tariff for wind power and set four types of wind power benchmark prices across the country. The fixed pricing mechanism ensured investor profits and thus promoted the development in the first few years [31]. The benchmark price is consisted of two components. The first is benchmark price of newly constructed de-sulfurized coal-fired power generation and the second is subsidy set by the central government and paid by means of surcharge in the retail tariff. Large-scale integration faces challenges under such a fixed pricing mechanism. The major reason is that the imbalance in the regional distribution of wind resources and power load centers creates the need for cross-provincial trade, and fixed pricing has blocked its implementation. A case study by [12] revealed that different fixed prices for different areas make it difficult and even impossible to trade wind power across the provinces.

5.2.2. Power Grid Companies

In light of the developmental pattern, the grid companies play an essential role in wind power development in China. However, the grid companies have complex attitude and their roles in developing wind power are complicated. Large-scale integration of wind power has imposed great challenges to the safe operation of power grids in many ways.

The intermittence character of wind power increases the difficulty of peak regulation. According to field statistics, the probability of anti-peak regulation for wind power in Northeast, West Inner Mongolia and Jilin grids is 60%, 57% and 56% respectively. Because of wind power, in Jilin power grid the period of enlarged peak valley load amount to 201 days in a year. Without sufficient generation capacity for peak regulation, the curtailment of wind power in Jilin and Inner Mongolia is an unavoidable choice.

Wind power also brings the difficulty of voltage control for power grids. With the large-scale integration of wind farms, the power grids become difficult to be controlled. According to field statistics, the 110-KV transmission lines must be operated at more than 113 KV voltages and the bus-bar voltage of 220-KV lines must be operated at more than 238 voltages to support regular system voltage in the Xinjiang power grid. In that way, it is very difficult to adjust the operation voltage. Besides, the operation of wind farms excessively relies on reactive compensation and it further limits the flexibility of grid operation. For instance, when reactive compensation equipment in Tara region of West Inner Mongolia was outage, the peak voltage of the 220-KV system increased to 257 KV.

The poor anti-disturbance capability of wind turbines influences the safe operation of the power grids. When there is small perturbation in the power systems, outrage of wind turbines will bring second impact on the power grid and result in amplification of the faults. Wind power will also increase the shock frequency to the power grids.

What is more important, additional investment for transmission enlarges the financial burden of the grid companies and pushes up their operation cost. On the opposite they can't gain extra revenue in the existing pricing mechanism. In a word, wind power imposes great safety and operation difficulties to the grid companies but brings no extra income for them.

6. Police Recommendations

6.1. Promoting Domestic Wind Turbines Manufacturing by Stronger Independent Innovation

Domestic wind turbine manufacturing can directly reduce the capital cost of wind farms and promote the commercialization of wind power. It can also break the import dependence of key components and core technologies and enhance the competiveness of domestic wind turbine manufacturers. Currently the industry policy pertinent to wind turbine manufacturing is mainly concerned with the cultivation of domestic manufacturing capability and the expansion of domestic market. In the future, the focus should be switched to independent innovation capability. Therefore scientific industry planning and technology development pathways should be formulated. A technology innovation system with companies playing leading roles, guided by market demand and co-funded by public and private money should be established. The following policies are suggested to the government:

- Provide fiscal support and preferential tax to those manufacturers conducting independent innovation and enhance the protection of intellectual property rights.
- Formulate and enforce industrial standards for wind turbines and improve the generality of wind turbine equipment.
- Strengthen the monitor on market concentration and implement necessary antitrust measures to curb it.
- Enhance the quality control of components production process and construct a qualified supply system for key components.

6.2. Establishing a Specialized Wind Resource Survey Institution and Providing Elaborate Wind Resource Mapping

At present, there is no national wind resource survey institution in China. Accordingly there is no professional service for the approval of wind farm projects. With the operation of such an institution, thorough survey of wind resource can be conducted and elaborate wind resource mapping can be provided for the project approval.

6.3. Developing Wind Power in both Concentrated and Distributed Ways and Encouraging Private Investment

To ease the difficulty of grid connection, it is comprehensible that China began with the way of concentrated development. However, basically wind power is distributed, so concentrated way won't work in the long run. Once distributed development is initiated, the next step is to lift the control on investment. The government should establish a level playground for all prospective investors.

6.4. Speeding up the Construction of Power Grids

To resolve the issue of grid connection, following suggestions are proposed:

- The government can set up a performance system to measure the grid companies.
- The government can provide subsidy to recover the additional cost of grid companies for integrating wind power.
- Speed up the construction of Smart Grids. Ultra-high voltage transmissions system can promote grid access of large-scale wind farms, while smart distribution network can promote the development of distributed wind power generation.

6.5. Optimizing the Dispatching of Power Grids

In China, under the current conditions, it is difficult for coal and CHP plants to conduct deep load regulation without technical retrofitting and reform of the dispatching code. The following ways can be used to solve the issue.

- Add more pumped storage hydropower or gas plants to improve the peak load regulation capability of the power grids.
- Provide incentive for those coal plants which provide peak load regulation service.
- Incorporate load regulation into the performance indexes of coal power plants and disclose the information in the annual SERC regulation report.

6.6. Attaching Importance to Energy Storage and Demand-Side Response

The massive development of wind power must be supported by large-scale energy storage. The introduction of energy storage facility can effectively reduce the shock of wind power on the power systems. In northeast or northwest China where wind resources are abundant and heating load during heating season is high, combination of wind farms with regional heating system can bring forth huge benefit. Electric heating or energy storage boiler can be viable solutions. Furthermore, demand side management can promote multiple utilization of wind power at valley period and a flexible pricing mechanism for wind power is needed to implement it.

CONCLUSION

Wind power is the mainstream green energy source with massive potential. With the publication and implementation of a set of energy policies, large-scale wind power development is taking off in China. In this chapter, with a perspective of supply chain, we try to figure out the panorama of China's wind power industry. Key stakeholders in the supply chain and their concerns are identified to understand the challenges of the industry. With a brief outline of the pertinent policies, status quo and issues of the supply chain are analyzed. Three remarks are made to conclude the chapter.

First, regarding the wind turbine manufacturing, the orientation of industry policy should be independent innovation capability, instead of the scale of domestic manufacturing. To foster innovation, a strict national technology standard system should be formulated to elevate the entry threshold. The government should keep a close eye on market competition behaviors and intervene with anti-trust measures at necessary time when vicious competition is detected.

Second, regarding the wind farm development, the orientation of the government should be a balanced planning with full consideration of the grid integration capacity and a level playground for all prospective investor. The development pattern of wind farm should also be amended. In addition to the current concentrated development of large-scale wind farms, distributed medium-and-small-scale wind farms directly balanced with local demand would also be encouraged to reduce the burden of grid integration. The development of smart grids will provide strong support for this new pattern.

Third, regarding the final utilization of wind power, grid companies should be properly motivated to integrate more wind power into the power grids. Formulation of practical technical codes for wind integration and dispatching rules friendly to wind power and other renewable energy will be a good start. Then proper allocation and recovery of the additional investment and operation costs should be settled by reform of the tariff mechanism. Separating the transmission and distribution price from the retail prices, and pricing the transmission and distribution service based not only on its cost but also on service quality, can provide strong incentive for grid companies, while the demonstration of distributed generation and microgrid will exert pressures on grid companies to improve their service quality.

REFERENCES

[1] Hu J. Presentation at Climate Change Summit Meeting in New York [online]. 2009[cited 2009 9 22]. Available from: *http://news.xinhuanet.com/world/2009-09/23/content_12098887.htm*

[2] Li J. (2012). *China Wind Power Outlook 2012*. Beijing: China Environment Science Press.

[3] Du Q; Xi X. Recommendations for facilitating the development of wind power industry in China. *China Soft Science*, 2001 10, 9-14.

[4] Yu J; Zhou H. Wind power development in China: the status and outlook. *Compressor Blower & Fan Technology*, 2006 6, 46-50.

[5] Kang C; Gao X; Rui X; Liu Y. The problems and measures in wind power industry in China. *China Energy*, 2006 28(6), 27-30.

[6] Shi P. Power transmission, rather than wind power generation is important. *Manufacturing Information Engineering of China*, 2011 12, 20.

[7] Xie G; Li Q; Wu Y. Research on comprehensive solution for coordinated development of wind power and grid. *Energy Technology and Economics*, 2011 23, 19-23.

[8] Wang Z; Qin H; Lewis J. China's wind power industry: Policy support, technological achievements, and emerging challenges. *Energy Policy*, 2012 51, 80-88.

[9] Zhang S; Li X. Large scale wind power integration in China: Analysis from a policy perspective. *Renewable and Sustainable Energy Reviews*, 2012 16(2), 1110-1115.

[10] Luo G; Zhi F; Zhang X. Inconsistencies between China's wind power development and grid planning: an institutional perspective. *Renewable Energy*, 2012 48, 52-56.

[11] Yang M; Patiño-Echeverri D; Yang F. Wind power generation in China: Understanding the mismatch between capacity and generation. *Renewable Energy*, 2012 41, 145-151.

[12] Zhao, X; Wang, F; and Wang M. Large-scale utilization of wind power in China: Obstacles of conflict between market and planning. *Energy Policy*, 2012 48, 222–232.

[13] Zhao Z; Yan H; Zuo J; Tian Y; Zillante G. A critical review of factors affecting the wind power generation industry in China. *Renewable and Sustainable Energy Reviews*, 2013 19, 499-508.

[14] Gu W. The research of large-scale isolated wind power system. *China Energy*, 2008 30(11), p14-17.

[15] Wang B. The measures and recommendation for wind power turbine generators. *China Electrical Equipment Industry*, 2009 1, 40-42.

[16] Si L; Meng W; and Wang Y. Study on wind power industry supply chain model based on circulation economy. *Ecological Economy*, 2010 11,162-165.

[17] Liu Y; Kokko A. Wind power in China: Policy and development challenges. *Energy Policy*, 2010 38(10), 5520-5529.

[18] Ling Y; Cai X. Exploitation and utilization of the wind power and its perspective in China. *Renewable and Sustainable Energy Reviews*, 2012 16(4), 2111-2117.

[19] Hu Z; Wang J; Byrne J; Kurdgelashvili L. Review of wind power tariff policies in China. *Energy Policy*, 2013 53, 41-50.

[20] Liao C; Jochem E; Zhang Y; Farid N. Wind power development and policies in China. *Renewable Energy*, 2010 35(9), 1879-1886.

[21] Zhao Z; Hu J; Zuo J. Performance of wind power industry development in China: A Diamond Model study. *Renewable Energy*,2009 34(12), 2883-2891.

[22] Kang J; Yuan J; Hu Z; Xu Y. Review on wind power development and relevant policies in China during the 11th Five-Year-Plan period. *Renewable and Sustainable Energy Reviews*, 2012 16(4), 1907-1915.

[23] Zhang S; Andrews-Speed P; Zhao X. Political and institutional analysis of the successes and failures of China's wind power policy. *Energy Policy*, 2013 56, 331-340.

[24] Wee H-M; Yang W-H; Chou C-W; Padilan M. Renewable energy supply chains, performance, application barriers, and strategies for further development. *Renewable and Sustainable Energy Reviews*, 2012 16(8), 5451-5465.

[25] Ru P; Zhi Q; Zhang F; Zhong X; Li J; Su J. Behind the development of technology: The transition of innovation modes in China's wind turbine manufacturing industry. *Energy Policy*, 2012 43, 58-69.

[26] China Wind Power Industry Net (CWPIN). List of wind turbine manufacturers and components suppliers [online]. 2010 [cited 2010 8 30]. Available from: *http://www.windchn.com/webinfo/wfview000117317.html*

[27] Mi C; Yuan X; Meng W. Competitive intelligence studies on wind power industry based on patent map technology. *Journal of Intelligence*, 2010 29(11), 39-43.

[28] State Electricity Regulation Commission. Annual Regulation Report on Wind and Solar Power Generation 2010[online]. 2011[cited 2013 12 15]. Available from: *http://www.fenglifadian.com/down/show-13.html*

[29] National Energy Administration (NEA). Notice on the relevant work for strengthening wind power integration and absorption[online].2012[cited 2012 6 1]. Available from: *http://www.nea.gov.cn/2012-06/01/c_131624884.htm*

[30] *World Wildlife Fund, 2010. 2010 rating of designed operational entities (DOEs) accredited under the Clean Development Mechanism (CDM)* (prepared by Lambert Schneider and Lennart Mohr).

[31] Li, J. Decarbonising power generation in China—is the answer blowing in the wind? *Renewable and Sustainable Energy Reviews,* 2010 14, 1154–1171.

APPENDIX

List of wind turbine, components and service companies in China

Company	Abbreviation
Goldwind Science & Technology Co., Ltd.	Goldwind
Sinovel Wind Group Co., Ltd.	Sinovel
Guodian United Power Technology Co., Ltd.	United Power
China Ming Yang Wind Power Group Ltd.	Mingyang
Dong Fang Turbine Co., Ltd.	DTC
XEMC Windpower Co., Ltd.	XEMC
Shanghai Electric Group Co., Ltd.	Shanghai Electric
China Creative Wind Energy Co., Ltd.	CCWE
CSR Zhuzhou Electric Locomotive Research Institute Co., Ltd.	CSR Zhuzhou
CSIC (Chongqing) Haizhuang Windpower Equipment Co., Ltd.	Haizhuang
Zhejiang Windey Co., Ltd.	Windey
Envision Energy Ltd.	Envision Energy
Ningxia Yinxing Energy Co., Ltd.	Yinxing Energy
SANY Group	SANY
CET XJ Windpower technology Company	XJ Wind Power
Huayi Electric Apparatus Group	HY
Zhonghang Hui Teng Windpower Equipment Co., Ltd.	HT
Lianyungang Zhongfu Lianzhong Composites Group Co., Ltd.	Zhongfu Lianzhong
Shanghai FRP Research Institute Co., Ltd.	SHFRP
Nanjing High-Speed & Accurate Gear Group Co., Ltd.	NGC
CN GPOWER Gearbox Co., Ltd.	CN GPOWER
Hangzhou Advance Gearbox Group Co., Ltd.	ADVANCE
Yongji Xinshisu Electric Equipment Co., Ltd.	YONGJI
Lanzhou Electric Corporation	LEC
Xiangtan Electric Manufacturing Group	XEM

Company	**Abbreviation**
Zhuzhou Times New Material Technology Co., LTD.	TMT
Shanghai Electric Group Shanghai Electric Motor Co., Ltd.	SEC
Dalian Tianyuan Electrical Machinery Co., Ltd.	Tianyuan Electric
DEC Dongfang Electric Machinery Co., Ltd.	DEC Dongfang
Shanghai Nanyang Electrical Machine Complex Co., LTD.	Nanyang Electric
CORONA Science & Technology Co., Ltd.	CORONA
Sungrow Power Supply Co., Ltd.	SUNGROW
NARI Group Corporation	NARI
China Tianyuan New Energy Technology Co., Ltd.	TIANYUAN
Shanghai Jiao Tong University	SJTU
Shanghai Electrical Automation R&D institute Ltd Inc	Shanghai Electrical Automation R&D institute
Shanghai Huiya Electronic Co., Ltd.	HUIYA
Shanghai Electric Wind Co., Ltd.	Shanghai Wind
Nan JiFeng Electric Equipment Co., Ltd.	NJF
Fang Lu consulting co., Ltd.	Fang Lu consulting
Tianjin Xin Yuan electric technology co., Ltd.	Xin Yuan electric
CASC-Acciona (joint venture by China Aerospace Corporation, Acciona and Inceisa)	Acciona

In: Renewable Energy in China
Editors: J. Yuan, G. Luo and J. Chen
ISBN: 978-1-63117-610-4

Chapter 3

SOLAR POWER INDUSTRY IN CHINA

Jianguo Chen *
School of Humanities & Social Sciences, North China Electric Power University,
People's Republic of China

ABSTRACT

China is endowed with abundant solar energy resources and its PV industry has made great progress recently. However, in contrast to the great potential as well as the development level of leading countries, China's solar power industry seriously lags behind and is confronted with lot of problems. This chapter is a review on the progress and challenge in China's solar power industry. Firstly, resources, technical condition, as well as the status of industry development are reviewed. Then the industry structure and the policy system for the industry are discussed. Next, the recently-announced planning on solar power development is presented and the market prospective is analyzed. Finally, barriers in the industry are identified and policy suggestions are proposed to address them. In the future, China must keep a close eye not only on the innovation of PV manufacturing, but also on the penetration of solar power generation. Regarding the energy policy on solar power, the priority should focus on final utilization, not simply on capacity installation. FIT tariff policy is important to stimulate investment in solar power, but the tariff must be dynamically reset to promote learning-by-doing.

Keywords: Solar power industry; Market prospective; China

1. INTRODUCTION

Solar energy is the most abundant, the most widely distributed, and inexhaustible renewable energy on the Earth. Solar photovoltaic(PV) has the characteristics of

* Corresponding author: Email: 13693090278@163.com.

modularization, easy installation and maintenance, and flexibility in utilization. Therefore, almost all national countries in the world attach great importance to solar energy, due to the increased pressures in energy resources and global climate change.

With advances in technology and decreases in costs, solar power has become near-commercial renewable energy technologies following the massive development and utilization of wind power. Total global PV installed capacity was 24GW in 2009 and it was over 100GW in 2012. Europe occupies the largest share of the total installed PV capacity, followed by China, the U.S. and Japan[1].

It could be traced back to 1997 since the Chinese government attaches great importance to solar power to solve the problem of electricity-access for household located in remote areas. The "China Brightness Project" was launched in such a background[2]. Since 2008, China began to promote solar power in a concentrated way. During 2008-2012, a series of policies have be formulated and implemented to boost the deployment of solar power in China. In 2013, distributed PV also became an important development direction of China's solar PV industry.

This chapter addresses the development of China's PV industry in a combined perspective of resource, technology and policy. Section 2 will discuss the resource of solar energy resources in China. Section 3 will review the current situation of the PV industry. Section 4 will discuss the institutional structure of the PV industry and its supply chain. Section 5 will discuss the policy evolution. Section 6 will present a prospective on the industry and the last section will present challenges and policy implications.

2. Resource Potential and Techniical Progress

2.1. Resource Conditions

China is located in the Northern Hemisphere, with a total territory of 9.6 million square kilometers. China is rich in solar energy resources, while the land area and buildings suitable for receiving solar light are also great. The Qinghai-Tibet Plateau, the Loess Plateau, northern Hebei plateau, Inner Mongolia Plateau and other areas rich in solar energy resources account for two-thirds of the total land area. The radiation and sunshine hours of Tibet, Qinghai, Xinjiang, Gansu, Ningxia, and Inner Mongolia plateau in China are the highest at 2800-3300 hours, with total annual radiation at 140-200 kcal/cm^2; the Sichuan Basin, Hubei and Hunan provinces and Shannxi are areas with low solar energy resources, with total annual radiation at 80-100 kcal/cm^2; eastern, southern and northeastern part of China is of the medium level, with the total annual radiation at 130 kcal/cm^2. The abundance and stability of the resources provide vast potential of solar power development in China.

At the end of 2010, there were a total of about 45 billion m^2 gross floor areas. The roof and the south facade were 500 million m^2, and 20% of the available area could install about 100 GW of solar PV system. China has about 1.2 million km^2 of desert, and if 5% is used for solar PV, the installation would reach 5000 GW, resulting in an annual generation of 6000TWh, roughly the predicted total generation of China in 2015[3].

2.2. Technology Conditions

The use of the solar energy is largely constrained by the technology conditions. The development in polysilicon production, solar cells, crystalline silicon solar cells, and silicon-based thin-film battery is very fast in China. By 2010, China's actual output of polysilicon reached 45,000 ton, which increases the self-sufficiency rate from 10% in 2007 to 50%. The conversion efficiency of the commercial crystalline silicon cell is approaching 19%. The conversion rate of Mono crystalline and Polycrystalline cell industry has reached 18.5% and 17.2%[4], respectively. Commercialization of silicon thin film battery reaches the maximum efficiency above 8% and localization rate of production equipment has increased to 70% at present.

In 2009, the Chinese government launched the "Golden Sun Demonstration Project" to boost the process of commercialization and technological progress of the PV products. In 2010, China's newly installed PV capacity was 5GW, while the cumulative installed capacity reached 8GW. Besides, the 500kW inverter for grid connection and other key equipment have realized domestic manufacturing [5]. With further development of smart grids, the technology progress will provide a solid foundation for large-scale development of solar power in China.

3. Development Status

China began to conduct the R&D on solar cells in 1958 and made some achievements in the early 1990s, when it was mainly applied in communications and industrial areas. In 1995, the PV power generation began to be applied in specific fields and remote areas, and the application system had been established. In 2000 China began to construct the demonstration systems of PV generation and connect them to the power grid. The rapid development of China's PV industry began in 2004, and in 2007 China has become the world's largest producer of PV cells. In 2010 the production of solar cell reached 13GW, and cell component increased to 10GW, accounting for 45% of world production. The solar cell production has ranked No.1 for five consecutive years in the world [4]. But the industry is suffered from a lot of issues, including the structure of the industry, the heavy reliance on overseas market, and others. In 2012, the solar PV industry suffered great difficulties, due to the anti-dumping investigation by the U.S. and EU.

3.1. PV Installation

The PV generation began to take shape in 1980s in China with just 280KW. It rapidly increased to 19MW in 1990 and to 70MW in 2005. The installation in the "Golden Sun demonstration project" is 696MW. In 2010 the newly-added capacity was 500MW and total installation was 0.9GW, which makes China the world's top 10 [6]. In 2011-2012 an obvious acceleration was witnessed. In 2012 new installation reached 4.5GW, while total capacity reached 8.3 GW [1]. However, with the rapid increase of the PV power generation, grid access can't keep pace with it. According to field statistics [7], at the end of 2011, about 0.8GW was not incorporated into the power grid and wasted [1].

3.2. Polysilicon Industry

Chinese polysilicon industry started in 1950s, and industrialization was realized in the middle of 1960s. In the 1970s, there were about 20 manufacturers. But because of backward technology level, serious environmental pollution, heavy material consumption and high cost, most of the enterprises stopped the production. In 1998, the government began to pay attention to solar power generation, and planned to build the first set of 3MW polycrystalline silicon cell and to kick off the demonstration project. In 2002, the world polysilicon production was 20350 tons, while China accounted for merely 0.25%. However, polysilicon production in China grew by an average annual growth rate of more than 100% during 2006-2010. By the end of 2011, the total production capacity in China has become the third largest in the world, with total capacity at 160000 tons, just after that of US and Germany [8]. By the end of 2012, China's polysilicon production capacity reached 190000 tons, but production dropped to 71000 tons. In 2012 China's polysilicon production capacity and output was the largest in the world, followed by the US (86000 tons), South Korea (57000 tons), Germany (55000 tons) and Japan (19000 tons) [4].

The development of polysilicon in China is reflected not only in the production, but also in technical progress. At present production line with capacity of 10000 tons has been completed and put into production in China. Cold hydrogenation technology has been popularized, and many manufacturers improve the comprehensive utilization rate through inverse disproportionation effect. Commercial silicon and polysilicon battery conversion efficiency has reached 18.5% and 17.2% respectively, entering the world advanced level. The cost of some leading manufacturers has approached $19/kg, the international advanced level [4]. The average material and energy consumption for producing 1 kg polysilicon were, 1.8-2.0 kg of industrial silicon, 1.8 kg of liquid chlorine, and 300-350 kWh of energy consumption; while in 2010 the numbers became: 1.3-1.4 kg industrial silicon, 1 kg of liquid chlorine, and 160-180 kWh energy consumption. Some backbone producer seven lowered the energy consumption to 130-150 KWh/kg. The use of polysilicon to produce PV cell fell from 11g/Wp in 2006 to 7-8g/Wp in 2010 [9].

Because of high dependence on international market to supplying raw materials and selling final products, in 2012 the industry experienced a heavy beat by the anti-dumping investigation, which further exacerbated the issue of overcapacity. According to "2013 China PV Development Report", by the end of 2012, there were only 8 enterprises still in operation and almost all other stopped production. This situation finally aroused the attention of the central government. As a quick response, the National Energy Administration (NEA) issued a specific planning to develop domestic PV generation in 2012. Soon after it, NEA issued a notice requiring provincial government to report the lists of demonstration projects.

3.3. PV Panel

From a point of view of production, China's domestic production of PV panel increased rapidly. The total production was only about 1000MW in 2007, and it increased to 2600MW in 2008, 4011 MW in 2009, 10800MW in 2010, 19800 MW in 2011, and 21000MW in 2012. Meanwhile, the production capacity also increased rapidly [4]. Regarding the type of the panel, crystal silicon accounted for 98%, while the rest of 400MW was thin film. For four consecutive years (2007-2010), China ranked No. in world PV panel production, and in 2010accounted for 50% of the global output [10].

From the point of view of technology level, the quality of PV panel product improves year by year, especially in the conversion efficiency. The performance of backbone producers improves by a significant margin, with single crystal silicon at 17-19%, polysilicon at 15-17%, and thin film at 6-8% [10].

From the point of view of market development, more than 90% of the PV panel products in China are exported. According to [4], in 2007exports reached \$3.49 billion, 2008 was \$5.83 billion, 2009 was \$20.19 billion and 2010 was \$20.2 billion. In 2011, the solar battery component export was about 16GW with a total sale of \$22.77 billion; in 2012 it was about 18GW, with a total sale of \$12.7 billion. Even though the export rose with large growth, the profits have dropped substantially.

4. The Institution Structure

4.1. PV Industry Chain in China

The PV industry chain includes six sectors, namely, silicon material, ingot casting (pull rods), slice, battery modules, battery, and application system. The upstream includes silicon material, silicon slice; midstream includes battery, battery modules, while the downstream is the application system (Table 3-1).

Table 3-1. PV industry chain

Upstream			Midstream		Downstream
crystalline silicon raw materials	Silicon rods / silicon ingots / Silicon slices		PV battery	PV modules	application system
polysilicon used to produce solar batteries	silicon rods	Silicon slices	Silicon battery	PV modules	PV generation system
	Polycrystalline silicon ingot	Polycrystalline silicon slices	Polycrystalline silicon panel		PV application products
Metallurgical grade silicon					
Silicon mine					

Table 3-2. Key producers in China's PV industry

Producer	Crystalline silicon	Silicon ingots	Silicon slices	PV battery	PV modules	PV system
YingLi Solar	●	●	●	●	■	▲
SUNTECH				▲	■	▲
Trina Solar		●	●	●	■	
LDK Solar	●	●	■		▲	
Hanwha-SolarOne		●	●	●	■	
JinkoSolar	●	●	▲	▲	■	
Canadian Solar Inc.			●	●	■	▲
DAQO Group	■	●	▲	●	▲	
GCL poly	■	●	▲	●		▲
ReneSolar	●	●		●	▲	
CEEGSST				■	▲	
JA Solar				■	▲	

● Product line, ■ Main sales, ▲ Secondary sales.

The main enterprises in the PV industry participate in almost all the sectors, and there are leading enterprises in the crystalline silicon, silicon ingots, slices, PV batteries, as well as PV modules and systems. Some large enterprises integrate all aspects of the industry chain. For example, YingLi Solar has product line in the crystalline silicon, silicon ingots, slices, PV batteries. The PV modules are its main sale products, and the PV systems are its secondary sale products (Table 3-2) [3].

4.2. Stakeholders of the Industry

According to the division of solar PV industry chain, the stakeholders are in the main fields of crystalline silicon materials, silicon rod/ingot/slice, PV batteries, PV modules, PV system and PV power integration and so on. According to the statistics of [4], there are 20-30 polycrystalline silicon companies, more than 60 silicon slice companies, more than 60 PV battery companies and more than 330 PV modules companies in China. However, international market also has fundamental impact on China's PV industry.

On December 7, 2012, the United States made the final ruling that the China's crystalline silicon PV panel and module products substantially endanger the development of solar PV industry in America. So the United States will impose punitive tariffs for these products. According to the standard ruled by U.S. Department of Commerce in October 2012, the United States will levy 13%-249.96% of anti-dumping duty and 14.78%-15.97% of anti-subsidy duty on China's crystalline silicon PV panels and modules. On September 6 2012, the EU also started the anti-dumping investigation on China's PV products. On June 5 2013, the EU made a preliminary ruling that EU will levy on Chinese PV products by 11.8% of the provisional anti-dumping duty from June 6 to August 6. If the EU and China failed to reach a solution on August 6, the anti-dumping rate will rise to 47.6% [11]. Therefore, foreign governments become indirect stakeholders of China's PV industry (Table 3-3).

Table 3-3. Stakeholders in China PV industry

Stakeholders	Property	Role	Concern
Domestic polycrystalline silicon producers	enterprise	Polycrystalline silicon raw material producers	profit
Foreign polycrystalline silicon producers	enterprise	Polycrystalline silicon raw material producers	profit
Domestic silicon rod / ingot / slice producers	enterprise	silicon rod / ingot / slice producers	profit
Foreign silicon rod / ingot / slice producers	enterprise	silicon rod / ingot / slice producers	profit
Domestic PV battery producers	enterprise	PV battery producers	profit
Foreign PV battery producers	enterprise	PV battery producers	profit
Domestic PV module producers	enterprise	PV module producers	profit
Foreign PV module producers	enterprise	PV module producers	profit
PV system developers	enterprise	PV system developers	profit
power grid	enterprise	Electricity transmission	profit
consumer	Families or enterprises	Electricity consumers	energy cost
The Chinese government	public sector	Planners, policy makers, inspirer, standard makers and anti-dumping executor	Environment, energy and industry development
The foreign government	public sector	Planners, policy makers, inspirer, standard makers and anti-dumping executor	Environment, energy and industry development

Data sources: collected and edited by the author.

5. Policy System

The regulation and policy system is also an important factor that affects the development of China's PV industry. This section reviews related national energy policy, R&D policy, electricity tariff policy and grid-connected policy, technical standard and investment policy.

5.1. Relevant National Energy Policy

The development of China's PV industry is tightly related to national energy policy on the development and utilization of renewable energy, including national laws and national planning system. For example, the "Electricity Law of China" promulgated in December 1995 stipulates that the state encourages and supports the utilization of solar energy in rural areas. On January of 2003, the State Council issued the "National Program for Sustainable Development in Early 21st Century", stipulating clear goals, principles, priorities and measures for China to implement sustainable development strategy in early 21st century. The "Law on Atmospheric Pollution Prevention" issued in April 2000 proposes that the state encourages and supports R&D on air pollution control, and utilization of clean energy technologies such as solar, wind, hydro, etc. (Table 3-4).

Table 3-4. The planning related to China's PV industry

Time	Title	Responsible agency	Key points
1986.12	The suggestion on strengthening the construction of rural energy from national economic council	state economic and trade commission	organize key demonstration projects in ten new technical fields, including development and utilization of solar energy and other new energy and renewable energy
1995.12	the Electricity Law	NPC(National People's Congress)	Encourage and support the utilization of solar energy, wind energy, geothermal energy, biomass energy and other energy in rural areas to increase rural electricity supply.
1999.1	The Law on Meteorology	NPC	Regulate that meteorological departments at different levels should make climate feasibility demonstration for urban planning, national key construction projects, major regional economic development projects and large development and utilization projects of climate resources energy, such as solar energy, wind energy etc.
2000.4	The Atmospheric Pollution Prevention Law	NPC(National People's Congress)	Encourage and support scientific and technological R&D on air pollution control, advance the development and utilization of clean energy such as solar, wind, hydro, etc.
2001.11	The 10th Five-Year Plan on Development of New and Renewable Energy	state economic and trade commission	Propose to build PV manufacturing system and to reduce the product cost with national public support, promote the commercialization of second generation PV battery and form a complete solar PV industry with well-equipped devices.
2003.1	National Program on Sustainable Development during Early 21st Century in China	the State Council	Focus on the development of clean energy, including natural gas, hydropower, renewable energy, new energy and so on.

Time	Title	Responsible agency	Key points
2005.2	the Renewable Energy Law	the State Council	Encourage companies and individuals to install and use solar energy system, such as, solar hot water system, solar heating and cooling system, PV generation systems etc.; Set up special funds to support R&D, standardization and demonstration projects for renewable energy.
2007.1	the Energy Conservation Law	NPC	Encourage the installation and utilization of solar energy and other renewable energy systems.
2007.8	Long-and Medium-Term Development Planning on Renewable Energy Resources	NDRC	Use household PV power generation system or small PV power station to solve the power supply problems for remote areas, including Tibet, Qinghai, Inner Mongolia, Xinjiang, Ningxia, Gansu, Yunnan and other provinces (autonomous regions and municipalities). Construct integrated roof solar power generation facilities. Build large-scale solar PV power station and solar thermal power plant.
2009.8	The National Action Program on Climate Change	NPC	Speed up the development of major technologies in response to climate change, including clean coal, renewable energy, nuclear energy and other low-carbon technologies.
2011.8	The Comprehensive Program on energy conservation and emissions reduction during the 12th FYP period	the State Council	Require that by 2015 non-fossil energy will account for 11.4% of the total energy consumption. Adjust measures to develop wind, solar, biomass energy, geothermal energy and other renewable energy sources.
2012.4	The 12th Five-Year Development Planning on Solar PV Industry	MOII	Put forward specific development goals and supporting policies for solar power during the 12th FYP period.
2012.4	The Special Plan on development of solar power technology during the 12thFYP Period	MOST	Put forward specific targets for solar power technology advancement and efficiency improvement.
2012.7	The 12th FYP on solar power generation	NEA	Promote sustained cost reduction and improve the competitiveness of solar power through market competition and scale development. Put forward a 20GW capacity installation target by 2015.
2013.7	Measures on promoting the sustainable development of solar generation	NEA	Increase the 2015 installation target to 35GW. Formulate a Feed-in-tariff (FIT) policy for concentrated solar power and subsidy policy for distributed solar power.

Data sources: collected and edited by the author.

Table 3-5. National PV power price

Resources areas	Benchmark price (yuan/KWh, taxinclusive)	Regions in resources areas
I	0.90	Ningxia, Qinghai hercynian, Jiayuguan, Wuwei, Dunhuang and Jinchang in Gansu province, Hami, Tacheng, Altay and Karamay in Xinjiang province and Inner Mongolia(except Chifeng, Tong Liao, Hinggan League, HulunBuir)
II	0.95	Beijing, Tianjin, Heilongjiang, Jilin, Liaoning, Sichuan, Yunnan, Chifeng, Tongliao, Hinggan League, HulunBuir in Inner Mongolia province, Chengteh, Zhangjiakou, Tangshan and Qinhuangdao in Hebei province, Datong, Shuozhou and Xinzhou in Shanxi province, Yulin and Yan'an in Shaanxi province, and other regions in Qinghai, Gansu and Xinjiang in included in I resources areas
III	1.00	other regions

5.2. Price Policy

PV generation now costs about three times more than coal power generation and is still in its early development in China and many other countries [12]. Therefore, it is a nearly a standardized measure to provide subsidy, in order to promote market penetration and technology learning. The Chinese government also has launched price policies in this way.

In January 2006 NDRC issues a tentative measure on renewable power generation pricing, stipulating that direct government pricing mechanism will be used for generation projects of solar energy, ocean energy and geothermal energy. Under this mechanism, full cost recovery and reasonable profit will be ensured. Franchising bidding is used by NDRC to find the competitive cost of PV projects. Then in 2007, NDRC issued another measure on the collection of renewable energy subsidy in term of surcharge in retail electricity tariff. Then in July of 2011, NDRC issued a new policy to simplify the management and pricing of PV projects. According to it, non-bidding solar power projects will receive a national unified benchmarking FIT price (1.15 yuan/KWh for projects approved before December 31, 2011 and 1 yuan/KWh for the projects approved hereafter.).

In August 2013, NDRC issued a new policy, which divides the whole nation into three resource areas in accordance with resource condition and construction costs and set the benchmark price. The difference of benchmark PV price and coal-fired price will be subsidized by the national renewable energy development fund (Table3-5) [13].

5.3. Policies on Solar Power Integration

In January 2011, the Ministry of Finance and the Ministry of Housing and Urban-Rural Development (MOHURD) issued "the Notice on the organization and implementation of demonstration projects for PV building integration", which promotes the integration from owners' side, and requires to accelerating the technology demonstration on micro-grid in public power grids. Then in October of 2012, China's largest grid operator, the State Grid Corporation of China (SGCC) officially released its guideline on grid integration of distributed PV generation. SGCC promises to provide favorable conditions for distributed PV integration and to invest in integration facilities with no charge to distributed PV generators. However, the scale and voltage level of distributed PV integration is narrowly stipulated by SGCC. On March 14, 2013, the former State Electricity Regulatory Commission issued its technical standard and safety requirements on PV projects. On August 29, 2013, NEA issued a regulation requiring that grid companies should carry out the planning and construction of power grids in line with national PV generation development planning and should adopt advanced technology such as smart grids as necessary to improve the capacity of PV integration.

5.4. Fiscal Subsidy Policies

Chinese government also provides fiscal and preferential tax policy to support PV industry. On May 30, 2006, the Ministry of Finance (MOF) issued the "Interim Measures on Special Fund Management for Development of Renewable Energy". Accordingly, the central government set up a special fund for renewable energy development, with a focus on wind energy, solar energy and ocean energy. On March 20, 2009, MOF issued a new notice to provide subsidy of 20000yuan/KWp for eligible solar power projects. The qualification will be determined on the degree of building integration and the technical level of the project. Then the annual subsidy standards will be adjusted according to the industrial development. On July 16, 2009, MOF, Ministry of Science and Technology and NEA issued two policies on Golden Sun Demonstration Project. In principle, Grid-connected PV power generation project will be subsidized with 50% of the total investment; for independent PV power generation system in remote areas, the subsidy rate will reach as high as 70%. Then on December 16, 2011, MOFlowered the subsidy to PV-building projects. For high integration project the subsidy will be 9000yuan/KWp, and for other projects it will be 7500yuan/KWp. Because the integration of PV into the power grid will incur extra costs to grid operators, if the extra cost is not properly recovered, grid company will be reluctant to integrate PV power. On March 14, 2012, MOF, NDRC and NEA jointly issued a policy, which provides subsidy to grid companies according to the lengths of the extra integration transmission lines. MOF also issued a policy in 2013 to provide 50% value-added tax exemption for solar power generators, which is quite similar to wind power. Table 3-6[4] provides an overview on subsidy policy for solar power.

6. Strategic Planning and Market Potential

In 2012, the Chinese government issued the "12th Five-Year-Plan on Solar Power Generation", the "12th FYP on Solar Industry" and the "Special Plan on Development of Solar Power Generation Technology during the 12th FYP Period".

Table 3-6. Subsidies policy for China's solar power generation

Subsidy type	Policy	Responsible agency	Time	Application	Subsidy methods	Fund source
investment subsidies	the Notice of Interim Measures on Financial Subsidies Fund Management for the Golden Sun Demonstration Project	MOF, MOST and NEA	2009.7.21	Solar PV power generation projects, industrialization of key technology, construction project of solar power generation capacity	According to the initial investment, on-grid project and independent system in remote areas without electricity enjoy subsidies of 50% and 70% respectively.	national renewable energy special fund
	the notice of Interim Measures on Financial Subsidies Fund Management for Renewable Energy Building	MOF, MOHURD	2009.3.23	application of PV building integration, formulation of the technical standards regulation, integration and promotion of common key technology	according to the installed capacity, the unit subsidy amount reduced year by year	national renewable energy special fund
price subsidies	the Notice on Policy to Perfect PV Generation Price	MOF, NDRC, NEA	2011.7.24	on-grid PV power projects	The fixed grid-connected price= benchmark price of desulfurization coal-fired electricity + subsidies	Renewable energy electricity price attached

It is interesting that the anti-dumping investigations by the US and EU unexpectedly leaded to Chinese government's adjustment on solar policy. It is evident that before 2012 solar power was not regarded as a large-scale renewable option in the eye of the government. However, the crisis of the domestic PV manufacturing industry finally pushed the government to formulate energy policy to help this industry. These plans pointed out guiding ideology and basic principles for solar power development. Specifically they proposed development targets, development and utilization layout and priorities of solar power generation, PV industry and solar power generation technology areas. In 2013 Chinese government fine-tuned the development targets and policies for solar power.

6.1. Strategic Goals

Specifically, the strategic goals on development of solar PV industry during the 12th FYP period cover both upstream and downstream targets, including: capacity of leading polysilicon companies attaining 50000 tons, and leading panel producers attaining 5GW (scale economy); recycled utilization rate of silicon tetrachloride, hydrochloric acid and hydrogen in polycrystalline silicon production attaining no less than 98.5%, 99% and 99% (material efficiency); conversion efficiency rate of silicon solar battery, polycrystalline silicon battery and amorphous silicon thin film battery reaching 21%, 19% and 12% respectively (technical progress); self-production rate of PV battery production equipment and auxiliary materials reaching no less than 80% (self-reliance); and most importantly, cutting the cost of PV generation, PV system and modules by a large extent (cost reduction) (Table 3-7) [10].

Table 3-8 [12] provides brief information on the strategic planning of China's solar power capacity during 2010-2020. It is clear that China aims to kick-off the market growth of solar power in the coming decade with rapid growth, with the hope of copying the successful experiences in developing wind power during 2006-2012.

Table 3-7. Strategic objectives of China's PV generation cost

Cost target	2015	2020
Cost of PV modules	7000 yuan/KW	5000 yuan/KW
Cost of PV systems	1.3*10^4 yuan/KW	1*10^4 yuan/KW
Cost of PV power generation	0.8 yuan/KW	0.6 yuan/KW

Table 3-8. strategic objective of China's solar power generation capacity (unit: GW)

Category	2010	2015		2020
		construction scale	key areas	construction scale
Solar power station	0.45	11		23
PV power station	0.45	10	Build a batch of on-grid PV power station in Qinghai, Gansu, Xinjiang, Inner Mongolia, Xizang, Ningxia, Shaanxi, Yunnan and suitable areas in North and Northeast China. Build a batch of PV power station combined with the large hydropower and wind power bases with complementary way.	20
Solar-thermal power stations	0	1	Carry out solar-thermal power demonstration projects in areas with good sunshine condition, wide available land and available water resource conditions.	3
Distributed PV	0.41	10	Build on-grid PV power generation system in east-central town industrial park, economic development zone and large public facilities for building roofs are relatively concentrated. Adopt independent PV power station or household PV system to deal with Electricity problem in non-grid areas. Expand the application of PV system in urban lighting and traffic signal systems.	27
total	0.86	21		50

Table 3-9. Investment estimation of China's solar power generation market

Category	Average cost (10^4yuan/KW)	2015 Installation (GW)	Investment estimation(billion yuan)	2020 Installation (GW)	Investment estimation(billion yuan)
Large power station	1	11	110	23	230
Distributed PV	1.5	10	150	27	405
total	-	21	260	50	635

6.2. The Market Prospect

The 12th FYP on solar power generation put forward that the newly added capacity of solar power will reach 10 GW and 11GW, for distributed PV and concentrated power generation, respectively by 2015. By 2020, the newly installation will increase to 23GW and 27 GW respectively. With an average investment cost of 10000 yuan/KW for large solar power and 15000yuan/KW for distributed solar power, the total investment for PV power generation is about 260 billion yuan in 2015 and 635 billion yuan in 2020 (Table 3-9) [12].

7. Barriers and Policy Options

Too much dependence on international market and insufficient development of domestic market is the key issue of China's PV industry. Excessive dependence on external markets is not conducive to sustained and healthy development of the industry. Since 2011, because of the negative impact of the global economic downturn, many European countries have slashed the investment and subsidies for solar power, leading to decline in global PV market. Therefore, now China switches to domestic market for the healthy development of its PV industry.

Technical upgrading is also urgently needed for the PV industry. The rapid expansion of production capacity of PV products and price competition result in poor technical development of the industry. There are many small-scale companies with low technical level. Poor products often disrupt the market and disorderly competition has occurred frequently. This industry upgrading requires active guidance from the government. Meanwhile, governments at all levels should not make inappropriate intervention to the industry. In particular, local governments are not supposed to blindly encourage constructing PV project to pursue economic growth and to use administrative measures to protect backward companies for local protectionism. Market mechanisms should be put in place to eliminate the inferior and promote resource integration.

China's domestic PV industry still lags behind in technology level. For instance, the key technologies of polycrystalline silicon lag behind the international advanced level; the high-grade equipment for crystalline silicon battery manufacturing is dependent on import; thin-film battery technology and equipment are also inferior to advanced levels. In terms of efficiency rate of crystalline silicon battery, the best practice has reached above 20%, while

China lags far behind. Regarding the performance of some key crystalline manufacturing equipment for silicon battery, there is a considerable gap behind the international advanced level. The standardization and testing system for battery, module and other PV components is yet to establish in China. Therefore, China needs to strengthen the technology R&D in PV industry by implementing innovation support policy.

Grid integration is also a key factor restricting the development of solar power. Regions with rich solar resources mainly locate in western and central China, and are far away from the power load center. Without adequate transmission lines, it is impossible to deliver solar power to final consumption. According to the research carried out by MOST in the first half of 2013, the curtailment of solar power in western China is as high as 40% [14]. Even some solar power stations with good grid-integration conditions only operate at 80% capacity rate. Therefore, the key for large-scale solar power development is grid-access and proper regulation on the grid companies' operation. A mechanism requiring that grid companies are responsible to the curtailment of renewable energy and must pay for the loss of renewable generators if there is curtailment, should be designed and put in operation in China, in so as to increase the responsibility of grid companies in renewable integration.

The incentive policies by Chinese government to promote the development of the solar PV industry are mainly subsidy policies, including subsidies both for capacity investment and electricity generation. Public subsidies in investment can reduce the investment risk and strongly stimulate the PV market, especially in its infancy period. But with the development of PV market, the drawbacks of investment subsidy show up. It is reported that many PV projects have not been put into real operation after receiving the investment subsidy. In this way the subsidies are actually wasted and don't attain the expected effect of promoting learning-by-doing. A feasible solution is to cancel the investment subsidy and redesign the FIT policy with similar sunset mechanism as Germany did.

CONCLUSION

China is endowed with abundant solar energy resources and its PV industry has made great progress. However, in contrast to the great potential as well as the development level of leading countries, China's solar power industry seriously lags behind and is confronted with lot of problems. In 2013 China issued a series of policies to support the development of domestic PV generation market, which radically change the situation of one-sided industry policy support to PV industry. In the future, China must keep a close eye not only on the innovation of PV manufacturing, but also on the penetration of solar power generation. Regarding the energy policy on solar power, the priority should focus on final utilization, not simply on capacity installation. FIT tariff policy is important to stimulate investment in solar power, but the tariff must be dynamically reset to reflect the effect of learning-by-doing.

REFERENCES

[1] Masson, G. et al. (2013). Global market outlook for photovoltaics 2013-2017. *European Photovoltaic Industry Association (EPIA)*, European Commission, May of 2013.

[2] Zhou, J. & Ma, S. (2001). Background and Plan of China Brightness Project. *Energy of China., 7.*

[3] Li, J. & Wang, S. (2011). *China Photovoltaic Development Report.* Beijing: China Environment Science Press, 2011.

[4] Li, J. & Wang, S. (2013). *China Photovoltaic Development Report.* Beijing: CREIA; 2013.

[5] China Power (website). Special Planning for Solar Power Technology Development during the "12th Five Year Plan" [online]. 2013. Available from: http://www.chinapower.com.cn/articleattachment/1198/art1198865.doc.

[6] Masson, G., Latour, M. & Biancardi, D. (2012). Global market outlook for photovoltaics until 2016. *European Photovoltaic Industry Association.*

[7] China Electricity Council. (2012). National Electric Power Industry Statistics Express (2011)[online]. Available from: http://www.cec.org.cn/xinxifabu/2012-01-13/78769.html.

[8] Polycrystalline Silicon Alliance. (2012). The 2011 Annual China Polysilicon Industry Research Report[online]. Available from: http://www.duojinggui.cn/?p=158.

[9] GreenPeace. (2012). Clean Production of Solar PV in China [online]. [cited 2012 3 27]. Available from: http://www.greenpeace.org/china/Global/china/publications/campaigns/climate-energy/2012/solar-clean-production.pdf.

[10] China Power (website). (2012). the 12th Five-Year Development Planning on solar PV industry [online]. Available from: http://www.chinapower.com.cn/article/1195/art1195941.asp.

[11] NetEase Finance. (2013). The anti-dumping investigation by the U.S. and EU [online]. Available from: http://money.163.com/special/guangfufanqingxiao/.

[12] China Power (website). (2012). the 12th Five-Year Development of Solar Generation Planning on [online]. Available from: http://www.chinapower.com.cn/newsarticle/1168/new1168279.asp.

[13] NRDC. (2013). The Notice on the Price Leverage to Promote the Healthy Development of the PV Industry [online]. Available from: http://www.china-nengyuan.com/news/49734.html.

[14] Economic Reference Daily. (2013). Grid regulatory upgrades to treat the curtailment of solar power[online]. http://www.cpnn.com.cn/zdzgtt/201312/t20131210_637668.html.

In: Renewable Energy in China
Editors: J. Yuan, G. Luo and J. Chen

ISBN: 978-1-63117-610-4

Chapter 4

THE DEVELOPMENT OF BIOMASS ENERGY INDUSTRY IN CHINA

Jianguo Chen *

School of Humanities and Social Sciences,
North China Electric Power University, People's Republic of China

ABSTRACT

China has made progress in biomass energy development, but its potential has yet to be fully deployed. In this chapter firstly the resource conditions of various biomass energy resources in China is analyzed. Then the development status of biomass is discussed. Policies relevant to biomass development in China are briefly discussed, with a perspective on the industry structure and institutional arrangement in this industry. Future planning for biomass energy in China is discussed. Finally the barriers to the planning are discussed and policy suggestions are proposed.

Keywords: Biomass energy, resource, planning, China

1. INTRODUCTION

Renewable energy has been developed for some important driving factors: the rising energy prices, in particular oil prices, energy security and the need to mitigate climate change effects. Biomass energy is not only in a pivotal position for energy conservation and emissions reduction but also can alleviate the pressure of climate change and optimize energy consumption structure. Biomass has the character of wide distribution and can be converted into gaseous, liquid or solid biofuels. Biomass energy is among the most promising green energy options due to the availability of huge quantities, reproducibility, diversification of energy products and significant comprehensive benefits. The development of biomass can improve the environment situation and increase energy supply, but contribute to rural

* Email: 13693090278@163.com.

development as well. The energy production from agricultural waste and animal manure can protect the environment from destruction. Experts estimated that the full utilization of current agricultural biomass resources in China can add about 500 million tons of standard coal, accounting for 24% of total production of primary energy in 2005 [1]. Through the utilization of biomass energy resources such as biogas, farmers can save money from energy conversion and get profits.

Biomass has a lot of advantages which attract high awareness of global countries and is considered as one of the most important resources to realize the national energy strategy. By the end of 2010, the global installed capacity of biomass power is more than 60GW [2]. The production and application of biogas, liquid biofuel and other biomass energy have attained a certain scale in Germany, Brazil, the United States and other countries. Today, Brazil has the world's largest-scale application of ethanol fuel accounted for 50% or more of gasoline consumption in the country. The annually power generation from biomass Combined Heat and Power (CHP) in Sweden reached about 100 TWh, represented 16.5% of national energy produced and 68.5% of total consumption for heating. The power generated by direct biomass combustion in Denmark will consume about 1.5 million tons of agriculture and forestry waste annually, providing 5% for the national electricity supply. The biomass energy is a significant part of the energy supply in The United States and Austria, representing 4% and 10% of the total primary energy supply [3]. Moreover, some developed countries also proposed ambitious plan for the development of biomass energy and put up goals in different stages. Some of them put forward medium and long-term development goals. In order to ensure the implementation of the development goals of biomass energy and biomass products, the United States Biomass Research and Development Technical Advisory Committee proposed *the American Biomass Technology Roadmap* on December 2002. According to the roadmap, the Commission pointed out the development direction for the future development of biomass technologies to Department of Energy, the Ministry of Agriculture, the Ministry of the Interior, the EPA, the National Science Foundation and the Office of Science and Technology Policy. The United States declares that the biofuels will take 20% of the total consumption of transportation fuels by 2020. *The Biofuels Promotion Law* promulgated by European Union in 2003 set the goals for liquid biofuel development in 2020. In 2005, European Union published *the Biomass Action Plan*, which covers all the development and utilization plans for biomass power generation, heat utilization of biomass energy and biological transportation fuel. In 2006, European Union also published *The EU biofuel development strategy*. These policies and plans play important role to promote the development of biomass energy for European Union.

The superiority of biomass energy also catches the attention of Chinese government. Through the policy and planning made by Chinese government, the bioenergy industry in China grew up rapidly, and expanded from small-scale to large-scale. *The Law of the People's Republic of China on Renewable Energy* promulgated in 2005 is the first legislation that encourages using clean and efficient ways to exploit and utilize biomass fuels, and the development of energy plants. *The outline of the 11th Five-Year Plan for National Economic and Social Development* (FYP) set in March 2006 required speeding up the development of biomass energy by supporting the development of crop straw and power generation from garbage incineration and landfill, constructing a batch of straw and wood power station, expanding the production capacity of biomass molded fuel (BMF), fuel ethanol and bio-diesel [4]. *The Long-and Medium-Term Development Planning on Renewable Energy Resources* published by National Development and Reform Commission (NDRC) in 2007 estimated that till 2020, the total installation capacity of biomass power generation will reach 30 GW, and annual consumption of BMF, biogas, biofuel ethanol and bio-diesel will reach 50 million tons, 44 billion m^3, 10 million tons and 2 million tons respectively [5]. *The Agriculture Biomass Industry Development Plan (2007-2015)* promulgated in 2007, analyzed the potential and current situation of the development of agricultural biomass energy in China, and indicated the development direction and priority. *The National Rural Biogas Construction Plan (2006-2010)* released by the Ministry of Agriculture in 2007 put up plans in three aspects, involved rural household biogas issues, large-and medium-sized biogas projects of large-scale farms and related technical and service supporting system. *National Medium and Long-Term Plan for Science and Technology Development (2006-2020)* promulgated in 2010, regarded biogas, solid and liquid biofuel and other biomass energy technology as key technologies for comprehensive development and utilization agricultural/forestry biomass. In 2012, *the 12th Five-Year Planning on Biomass Energy* issued by NDRC proposed that in 2015, the use of biomass energy should surpass 50 million tons of standard coal annually. *The Special Plan on Development of Biomass Energy Technology during the 12th FYP Period* published by the Ministry of Science and Technology in 2012 proposed to focus on the collection and storage of the traditional biomass resources, cultivation of the new biomass resource and conversion from biomass to gas, liquid and solid fuels and thermal energy efficiently, and put forward general idea, development goals and major tasks during the 12th FYP period.

The development of biomass energy not only depends on technology, but also involves the biomass resources, the industrial institutional structure and policy system, etc. Biomass resources are the physical basis and prerequisite condition of the development of biomass energy. The industrial institutional structure is the interest relation among the various parts of biomass energy industry chain. Policy system also is highly relevant to the incentive structure of the industry. The experience of developed countries shows that at the initial stage, the production cost, supporting infrastructure, and related technologies are difficult to compete with that of traditional energy sources. Policy support and institutional mechanisms are the main driving force for the science and technology development of biomass energy industry [6]. With the huge value in biomass sector, it is of great significance to realize the renewable energy strategy and the development of biomass energy industry in China.

The remainder of this chapter is organized as follows. Section 2 aims at analyzing in depth about the available amount of the biomass resources, which are the premise and basis of the development of biomass energy. Section 3 basically introduces the institutional structure of the biomass energy industry in China and related stakeholders by integrating the biomass industry chain and process. Section 4 investigates the policy issues on biomass energy, and analyzes the incentive structure that pushed the development of biomass energy in China. Section 5 presents current situation of China's biomass energy industry based on the resources and the institutional structure of the policy system. Section 6 analyzes the obstacles restricting the development of biomass energy industry in China through reviewing and evaluating the representing performance. Section 7 proposes policy recommendations to deal with the constraints and explore development prospects and the last section is conclusion.

2. Resources Conditions for the Development of Biomass Industry in China

Resources are the premise for the development of biomass energy industry. Therefore, the analysis of the resources of biomass energy in China is not only an explanation for the past achievements of biomass energy in China, but also an indication of the future development potential. China is abundant with biomass resources. The deployment patterns are various, including power generation, biogas supply, liquid biofuel, solid fuel and so on. In this section biomass resources including crop straw, poultry and livestock manure, forest biomass, municipal solid waste and energy plant are reviewed. However, industrial organic waste and algae microorganisms are not included in our analysis.

2.1. Agricultural Residues

There are abundant and a wide range of agricultural biomass resources in China, mainly including crop straw, livestock and poultry manure, by-products of agricultural product and energy plant, etc. The development potential is huge. According to investigation statistics, the quantity of national theoretical resource of straw stalk in 2010 including maize, rice, wheat, cotton, oil-bearing crops are 840 million tons and about 700 million tons can be collected. The resources are mainly distributed in 13 grain-producing provinces in the North China Plain and the Middle-lower Yangtze Plain and the Northeast Plain. Among them, there are about 211 million tons straw, 154 million tons wheat straw, 273 million tons maize stalk, 26 million tons cotton stalk, 37 million tons oil-bearing crops straw (mainly the rape and peanuts), 28 million tons legumes straw and 23 million tons potato stalk (table 1) [7]. In 2010, the consumption of crop straw reached about 500 million tons with 70.6% of the comprehensive utilization rate. Among the consumption, about 218 million tons crop straw used as fodder, accounting for 31.9%; about 107 million tons used as fertilizer, accounting for 15.6%; about 18 million tons used as planting edible fungus ground mass, accounting for 2.6%; about 18 million tons used as raw materials for artificial board, paper and other industrial material, accounting for 2.6%; and about 122 million tons used as fuel (including traditional cooking and heating, straw energy-oriented usage), accounting for 17.8%. The comprehensive utilization of crop straw is significant [7].

Table 4-1. The amount of field crop straw in China in 2010 (unit: Mt)

straw	wheat straw	maize stalk	cotton stalk	oil-bearing crops straw	legumes straw	potato stalk
211	154	273	26	37	28	23

According to the 12th FYP on Biomass Energy, in China there are about 3.4 trillion tons crop straws that can be possibly used, equivalent to 170 million tons of standard coal equivalent (sce). 8 million tons have been used, equivalent to 4 million tons of sce. And 332 million tons are available, equivalent to 166 million tons of sce (table 2) [2].

The agricultural and industry by-products include rice husk, maize cob, bagasse, etc. Most of them are from the food processing factory, packinghouse, sugar refinery and brewery. With the benefit of enormous production, centralized production and easy collection, the total amount of by-products production were more than 100 million tons in 2005, equivalent to 31-67 million tons of sce in maximum utilization [1]. According to the 12th FYP on Biomass Energy, there are about 120 million tons rice husk and bagasse residues produced annually in China, and about 60 million tons can be used for energy utilization per year.

Table 4-2. China's agricultural residue resources (unit: Mt)

Source of resources	Available resources		Used resources		The rest of the available resources	
	Quantity	Standard Coal Equivalent	Quantity	Standard Coal Equivalent	quantity	Standard Coal Equivalent
crop straw	340	170	8	4	332	166
Agricultural products processing residues	60	30	2	1	58	29

2.2. Forestry Residue Resources

At present, the acreage of existing forest land in China is about 300 million hectares. The acreage of existing natural forest is about 200 million hectares with forestry stock of 13.7 billion m^3. The acreage of artificial forest is about 60 million hectares with the accumulative volume of 1.96 billion m^3. The potential capacity of forestry biomass resources is about 18 billion tons. The existing forestry resources which can be used for energy are primarily wood resources, woody oil and starch plants.

The kinds of forestry wastes that can be utilized as biomass resources are mostly firewood forests, Bamboo residues, bush, residues from cross cutting and wood processing, economic forests and pruning branches from urban greening, etc. The existing firewood forest in China are more than 170 hectares with accumulative volume of 39 million m^3, mainly distributed in Yunnan, Liaoning, Shaanxi, Hubei, Sichuan and other 15 provinces, accounting for 97% of the total acreage and volume of national firewood forests. The forestry harvest volume is about 250 million m^3, which can produce about 110 million tons of cutting and building materials residues. The existing bush with a total acreage of 5300 hectares can collect 100 million tons of woody fuel annually. There are more than 7000 hectares of sapling forest which supply 20 ~ 40 million tons of raw materials from tending and pruning (table 3) [3]. Also, pruning branches from economic forests and urban greening can provide some raw materials. The potential quantity of the forest resources which can be used as wood energy is about 350 million tons in China, equivalent to 200 million tons of sce by utmost usage [3].

Table 4-3. Amount of residue of forestry resources in China

Species	Acreage (hectares)	Volume (m^3)	Annual deforestation (m^3)	Annual residue (Mt)
firewood forest	1.7 million	3.9 million	250 million	110
bush	53 million			100
sapling forest				20~40

Overall, according to the statistics, the forestry residue in China is 350 million tons, equivalent to 200 million tons of sce. Only 3 million tons residue have been used, equivalent to 1.7 million tons of sce, while there are 347 million tons residues available, equivalent to 189.3 million tons of sce (table 4) [2].

Table 4-4. Total amount of forestry residue resources in China and its energy potential (unit: Mt)

Source of resources	Available residue		Used residue		Rest residue	
	Quantity	Standard Coal Equivalent	Quantity	Standard Coal Equivalent	Quantity	Standard Coal Equivalent
Forestry residue	350	200	3	1.7	347	198.3

2.3. Livestock and Poultry Manure

Agricultural biogas resources mainly come from the centralized feeding farms. The mainly resources are manure from swine, cattle and poultry. According to the statistics from Department of Agriculture, the growth rate of stock farming kept at more than 8% in recent years. By the end of 2011, there are 119.662 million large livestock in China, 750.027 million swine and sheep. Their manure is the main source of biogas resources (table 5)[8].

Table 4-5. The feeding quantity of livestock in China by the end of 2011(Million)

Large livestock								Sheep stock	
quantity	cattle	horse	donkey	mule	camel	Pig slaughter	Pig stock	goat	sheep
119.662	103.605	6.709	6.478	2.598	0.273	661.703	467.669	142.742	139.615

According to statistics from the report for present situation of biogas industry development in China, there are 1.48732592 billion tons manure in China in 2009, 1.921057 billion tons urine (table 6).

Table 4-6. The amount of faeces from poultry and livestock in 2009

	Stock	Daily manure (kg per)	Daily urine (kg per)	Manure (Mt)	Urine (Mt)
cattle	107.265 million	15	25	587.27588	978.79313
swine	645.386 million	3	4	706.69767	942.26356
chicken				193.35237	0
total				1,487.32592	1,921.057

Note: There is no quantity of chicken in statistical yearbook; the data in this report are calculated based on the estimate that chicken manure accounts for 13% of the total livestock and poultry manure. The daily manure and urine is from empirical data.

Overall, according to statistics of the 12th FYP on Biomass Energy, there are 840 million tons poultry and livestock manure resources in China, equivalent to 28 million tons of sce. There are 300 million tons manure resources in usage, equivalent to 10 million tons of sce, while 540 million tons manure resources are available, equivalent to 18 million tons of sce (table 7) [2].

Table 4-7. Total amount of poultry and livestock manure in China and its energy potential (Unit: Mt)

Source of resources	Available resources		Used resource		Rest resources	
	Quantity	Standard Coal Equivalent	Quantity	Standard Coal Equivalent	Quantity	Standard Coal Equivalent
Livestock and poultry manure	840	28	300	10	540	18

2.4. Municipal Domestic Organic Wastes

According to the data of *China Statistical Yearbook 2012*, the municipal domestic organic wastes in China are 163.9528 million tons. 50% of them can be used as the materials for power generation from incineration and landfill, and the energy potential was equivalent to 12 million tons sce. Kitchen wastes also can be used as raw materials of bio-diesel, which is amount about 3 million tons sce per year. Annual sludge production of urban sewage treatment plant is about 30 million tons, of which about 50% can be used for energy production. More than 20 industries, such as alcohol, sugar, wine etc. produce about 4.35 billion tons of organic wastewater and 950 million tons of offscum each year, which can be converted into about 30 billion m^3 biogas. The manure resources from large-scale livestock and poultry farm are about 840 million tons every year, which potentially produce about 40 billion m^3 biogas [2].

2.5. Energy Plant

The most widely planted energy crops in China include sweet sorghum, cassava, sweet potatoes, and rapeseed. Sweet sorghum, cassava and sweet potatoes can extract carbohydrate to produce bioethanol. Rapeseed is the only bio-diesel raw material people concerned. The fuel can be used without refitting engine and also can be mixed with diesel. According to the statistics from the 12th FYP on Biomass Energy, more than 30 species energy crops are suitable to be planted in China, including oil palm, jatropha curcas, lacquer tree, sorbifolia, pistacia chinensis, sapium, sweet sorghum and so on, which can fulfill the need for producing 50 million tons of liquid fuel. According to statistical data, China is the world's large rapeseed, sweet sorghum and sweet potato producer. The planting acreage of rapeseed in China is 7.347 million hectares in 2011 with 13.426 million tons output. The planting acreage of sorghum is 0.5 million hectares with 2.051 million tons of output. According to *the Study of the Technology Roadmap of Biomass Energy in China*, available biomass liquid fuel in China is equivalent to 2.71 million sce in 2008 (table 8) [9].

Table 4-8. The amount of the 2008 biomass liquid fuel resources and the amount can be obtained (Unit: Mt)

Species	Total quantity	Available quantity	Standard Coal Equivalent
cassava	11	1.4	0.21
sugar grass	2.5	2.5	0.16
waste oils and fats	2	2	1.71
Woody oil fruit	2.2	1.7	0.62
total			2.71

3. The Current Situation of the Development of Biomass Energy Industry in China

The exploitation and utilization of biological energy in China began in the 1970s. But at present, the industry is still in the early stage of development due to slow development rate. The main areas of technology development focused on fuel ethanol, bio-diesel, biogas, biomass power generation/heating, straw briquette fuel, etc. In 2010, the usage of biomass energy (not including direct burning firewood using traditional way) is equivalent to about 24 million tons of sce [2].

3.1. Biology Liquid Fuel

3.1.1. Biofuel Ethanol

From 2001, the national government appointed experimental sites for usage of ethanol alcohol gasoline for vehicles in Henan, Heilongjiang, Jilin, Anhui and other nine zones. During the 10th FYP period, there are four fuel ethanol demonstration projects, viz., Jilin fuel ethanol co., Ltd., Henan Tianguan Group Co., Ltd., Anhui Bbca Biochemical Stock Co., Ltd.

and Heilongjiang Huarun ethanol co., Ltd., which are approved to use the rotten rice to produce biofuel ethanol. In recent years, several provinces and cities start to expand ethanol application and have achieved obvious results. The pilot provinces with 10% ethanol content increased from four to nine provinces. With the expansion of the pilot projects, the sales of fuel ethanol grow rapidly in China. The production capacity in 2009 reached 1.93 million tons annually with the actual output of 1.732 million tons [6]. In 2011, China had been the third largest producer of ethanol with 1.9376 million tons of fuel ethanol (table 9) [10].

Table 4-9. The output of fuel ethanol in China (unit: Mt)

year	2005	2006	2007	2008	2009	2010	2011
output	1.02	1.07	1.26	1.5	1.73	1.87	1.93

3.1.2. Bio-Diesel

According to statistics, the annual production capacity of bio-diesel industry in China is more than 3 million tons by the end of 2007. According to incomplete statistics, there are 26 bio-diesel companies with production capacity of 10,000 tons or less, 13 companies with the capacity of less than 50,000 tons, 7 companies with the capacity between 50,000 to 100,000 tons, and 6 companies with the capacity more than 100,000 tons. Although the existing and planned capability of bio-diesel energy in China is considerable, the industry development lags behind with a small actual annual production of only around 0.4-0.5 million tons and low capacity utilization rate at 20% -25%. Most of the products are not sale as refined petroleum but as chemical products [11].

3.2. Biogas

China is one of the earliest countries on research and development of biogas and has the advanced biogas technologies. During the period of 6th and 7th FYP, the government took the biogas technology development as a key point in national scientific research plan. In 8th and 9th FYP period, the government built an efficient agriculture production mode with multi-level use of substance and rational energy flow. The ministry of agriculture put forward "environmental energy project" and "gaining profit from ecologicalization project" in 1999. With the support of relevant national policy, the amount of biogas digester in rural household raised rapidly at the growth rate of 4 million blocks per year. By the end of 2009, there were 35.0703 million household biogas digesters (below 12 m^3) in rural areas of China. 32.3024 million rural families use biogas with an annual output of 12.4 billion m^3 of biogas. The number of biogas projects increased dramatically also. By 2009, there were totally 56,856 biogas projects with annual biogas output of 917.2809 million m^3. There were also 322 large-sized biogas projects (with digester volume more than 500 m^3) treating industrial waste with the total volume of 561,200 m^3 and annual biogas output of 152.3582 million m^3, providing gas for 367,200 users. These projects with a total capacity of 9,051.5 KW, achieve electricity generation of 49.636 million KWh annually. There are 56,534 biogas projects treating agricultural waste, these projects provide 764.9217 million m^3 of biogas and meet the need of 971,862 gas users. Among them: 3,717 are large-sized biogas projects (the digester volume is more than 500 m^3); 18,853 are medium-sized projects (the digester volume is between 50 m^3

to 500 m^3); 33,964 are small-sized biogas projects (with digester volume less than 50 m^3). These projects amount a total installation of 33MW and achieve electricity generation of 102.89 GWh [6].

3.3. Biomass Power Generation

In 2005, National Development and Reform Commission (NDRC) approved three straw power generation demonstration projects, located in Jiangsu, Shandong and Hebei province respectively. Since then, biomass power generation project in China started to grow up. By the end of 2005, the total installed capacity of biomass energy reached about 2 GW, with 1.7 GW from bagasse, 0.2 GW from garbage, and the rest from methane and the waste of agriculture and forestry such as rice husk etc. [5]. By the end of 2010, the installed capacity of biomass energy increased to 5.5 GW, with 1.9 GW from agriculture and forestry biomass, 1.7 GW from garbage power generation, 1.7 GW from bagasse, and 0.2 GW from biogas and other biomass energy. With annual power generation output of 20 TWh, the corresponding annual consumption of agriculture and forestry residues are about 10 million tons, which totally increase about CNY3 billion to the income of farmers. For the manufacturing aspect, China has mastered the biomass power generation technology with high temperature and high pressure [12].

3.4. BMF

According to *the 12th FYP on Biomass Energy*, in 2010, the production of BMF is about 3 million tons in China, mainly used for rural residents and used as fuel of urban heating boiler and raw materials of biomass charcoal.

4. Industry System Analysis for Biomass Energy Development in China

There are several factors driving the development of the biomass energy industry in China. In addition to the resource condition, another important factor is institutional structure, which provides compatible incentive mechanism to related stakeholders.

4.1. The Stakeholders of the Biomass Industry in China

The government is the primary actors in the development of biomass energy in China, including the central government and local governments at all levels. They focus on how to promote the development of biomass energy and solve the problem of energy structure and environment pressure. The main role of the central government is making planning, setting goals and promulgating corresponding incentive policies to guide and standardize the development of biomass energy in China. The governments involved in the biomass energy development are NDRC, the national energy administration (NEA) and the ministry of agriculture.

The local governments are mainly responsible for policy implementation, organization, publicity, mobilization, guidance, demonstration and technology promotion. The local governments involve the energy agency of the prefecture and township government. In addition, grassroots autonomous organizations of the villagers committee also impact the effect of mobilization, propaganda and organization. Technology promotion agency and biogas system technical personnel are also the important stakeholders in the process of biomass energy development.

Peasant household also play an important role in the process of biomass development as both supplier and consumer of biomass raw materials. They provide agricultural waste for the production of biomass. There are two possible measures for peasant household to offer raw materials. The first is putting agricultural waste into household biogas digester to produce biogas for cooking and heating, and use organic fertilizer in their own fields; the other is selling waste to large or medium-sized biogas project [13]. Livestock and poultry farms also play the dual role of raw material suppliers and consumers. They could put poultry and animal feces into farm digester to produce and sell biogas. They can also sell animal manure to related biogas producers.

Table 4-10. Stakeholders in the development of biomass energy industry in China

Stakeholders	**property**	**roles**	**interests**
central government	public	decision-maker, designer and incentive providers	energy supply, clean energy
The local government	public	executive and implementer	the local environment and economic development
peasant household	private	raw materials suppliers and energy products consumer	raw material prices and energy costs
Livestock and poultry farms	private company	raw materials suppliers and energy products consumer	raw material prices and energy costs
Biomass energy producers	private company	the producers of biomass energy	profit
Biomass energy equipment manufacturers	private company	equipment manufacturer or facility builder	profit
Biomass energy product delivery agency	private company	transport biomass energy products	profit
Biomass intermediaries	private company	biomass collector, storage, transporter and distributor	profit
Biomass energy users	mixture	consumer	energy cost and supply adequacy

Data sources: collected and compiled by the author.

Biomass energy producers, mainly include straw briquette fuel companies, fuel ethanol producers, biomass diesel oil producers, biogas gas suppliers, and biomass generation companies. Biomass energy equipment manufacturers, who produce special equipment for biomass energy, mainly provide pretreatment equipment and boiler for direct combustion, power generation and so on. Delivery agency mainly includes pipeline for biomass gas and biogas, distributors and transporters for BMF, power grid for distributing biomass power generation, and so on. Biomass intermediaries are mainly responsible for collection, storage and transportation for biomass raw materials, and provide the materials to biomass energy producers. The property, roles and interests of the actors in the biomass industry in China are shown in table 10.

4.2. Institutional Structure Related to Biomass Energy in China

The relationship between the rights and duties of the stakeholders discussed above constitutes the institutional structure of Biomass energy industry in China. The institutional structure mainly regulates the supply of raw materials, technology and equipment R&D, the production, distribution and consumption of the product. Preferential institutional system concluding fiscal taxation, product price, marketing, import and export, electricity price subsidies, special reward, credit guarantee and industrial development special funds etc., together form a regime structure to support biomass energy development in China (Fig. 4-1).

5. Policy Evolution of Chinese Biomass Energy Industry

Biomass energy development in China relies not only on the resources and institutional structure, but also on national policies system. The system is composed of sector planning, laws/regulations, and incentive support including preferential finance and taxation.

5.1. Planning for Biomass Energy

Planning associated with biomass energy plays a strategic guiding role to set up goals and point out development priority. The planning related to the development of biomass energy in China is shown in table 11.

Table 4-11. Planning system of biomass energy in China

Time	Planning	Key points
2001.10	The 10th Five-Year Plan on Development of New and Renewable Energy	Focuses on developing large and medium-sized biogas projects, which dispose high concentrated organic wastewater with anaerobic digestion technology, and cultivating related equipment manufacturing.
2007.4	The 11th Five-Year Plan on Energy Development	Regards biomass liquid fuel, coal chemical and coal-based liquid fuel technology as key points for the development in oil replacement project.
2007.7	The Plan on Agriculture Biomass Industry Development (2007-2015)	Sets the strategic targets and key tasks for the utilization of rural biogas straw energy and energy plant.

Time	Planning	Key points
2007.9	Long-and Medium-Term Development Planning on Renewable Energy Resources	Sets the targets as by 2010, the total installation capacity for biomass power generation will be 5.5million KW, and it will achieved 30 million KW till 2020.
2006.2	National medium and long-term plan for science and technology development（2006-2020）	Regards biogas, biomass liquid/solid fuel and other biomass energy as key technologies for comprehensive exploitation and utilization of agricultural\forestry biomass.
2007.4	National Plan for Rural Biogas Construction (2007-2010)	Estimates that by 2010, about 40 million families living in the countryside will use biogas as the main domestic fuel, and about 15.4 billion m^3 biogas will be consumed per year, equal to 2.42 million tons of standard coal and 140 million acres of the stock volume of forestland.
2007.4	Development Scheme for National Rural Biogas Service System	Proposes that during the 11th Five-Year Plan period, the technical network of biogas will cover more that 70% rural areas in China, forming an efficient technical service system.
2012.5	The Special Plan on development of biomass energy technology during the 12th Five-Year Period	Lists the key tasks for the development of biomass energy technology, including: the production and efficient utilization usage of biogas, the development of advanced liquid biofuel, the development of standardized sets of equipment development for BMF production and application, the cultivation of high-biomass energy plant and the production of bio-chemical products, the breeding and biological refining of energy microalgae, power generation from efficient combustion of biomass and the new gasification technology.
2013.5	National Plan for Forestry Biomass Development (2011-2020)	Forms the industrial system of forestry biomass contains cultivation, production, transformation and application by 2020. Build 16.78 million hectares energy forest and annual consumption of forestry biomass will be more than 20 million tons of standard coal by 2020.
2012.7	the 12th Five-Year Development Planning on Biomass Energy	Proposes that by the year 2015, the annual consumption of biomass energy should be more than 50 million tons of standard coal.
2012.8	the 12th Five-Year Development Planning on Renewable Energy Resources	Points out that in 2015, the consumption of biomass energy should be equal to50 million tons of standard coal.

Data sources: Collected and edited by the author.

5.2. Laws and Regulations on Biomass Energy

The Law of the People's Republic of China on Renewable Energy was implemented on Jan.1 2006. This legislation relates with bioenergy in the following points: advocating clean and efficient exploitation/utilization of biofuels, planting of energy crops; encouraging production and utilization of liquid biofuel; promoting the utilization of biofuels conversion according to local circumstances; etc. Then, supporting regulations include Management

Measures for Price and Sharing of Expenses for Electricity Generation from Renewable Energy (Jan, 2006) and Management Regulations for Electricity Generation from Renewable Energy (Feb, 2006). The regulation further defines the approval and management process for biomass power generation projects and the related preferential tariff. The law and regulations also encourage enterprises entering the renewable energy industry and provide guidance and standard. Besides, policies in price subsidy and preferential financial taxation are strong guarantee for the development of bioenergy industry (Table 12).

Table 4-12. Laws and regulations for biomass energy in China

Time	Name	Departments	Contents
2005.2	the Law of the People's Republic of China on Renewable Energy Resources	National People's Congress (NPC)	States that the state advocates clean and efficient exploitation/utilization of biofuels, as well as planting of energy crops. Proposes encouragement and support to the exploitation and utilization of renewable energy in rural areas.
2005.11	The Guidance Catalogue for Development of Renewable Energy Industry	**NDRC**	Promotes commercialization and mass utilization of large-and medium-sized biogas projects; Regards power generation from municipal domestic waste as basic commercial projects; Regards power generation from direct combustion, BMF and energy plant as demonstration and promotion projects; The projects above would get the advantage from technology research, project demonstration, financial taxation, product price, marketing, import and export etc.
2006.1	Management Regulations for Electricity Generation from Renewable Energy	NDRC	Stipulates that power grid enterprises shall purchase power generated by renewable energy projects within its coverage in full amount.
2006.1	Tentative Management Measures for Price and Sharing of Expenses for Electricity Generation from Renewable Energy	NDRC	Determines the subsidy price of biomass power generation project is CNY0.25/KWh. The project would get the subsidy for 15 years from operation. The subsidy will be canceled when the project has operated for 15 years.
2006.5	Interim Measures on Special Fund Management for Development of Renewable Energy	Ministry of Finance	Regulates management of the special fund for renewable energy development by "Free financing" and "Loan discount", and determines fuel ethanol and bio-diesel as among key supported works.

Time	Name	Departments	Contents
2006.9	Implementation Opinions on The Development of Financial/taxation Support Policies of Biomass Energy and Bio-chemical Projects	Ministry of Finance, NDRC, Ministry of Agriculture, SAT(State Administration of Taxation, SFA(State Forestry Administration)	Provides financial/ taxation support to biomass energy and bio-chemical industry by flexible loss subsidy, raw material base subsidy, demonstration subsidy and taxation favor etc. and determines the relevant implement method.
2007.7	Interim Measures on Encourage Fund Management for Non-grain Instruction for Development of Biomass Energy and Bio-chemical Industry	Ministry of Finance	Sets non-grain instruction encourage fund to support production of ethanol from straw, sweet sorghum and potato, production of bio-diesel, and co-production of bio-chemistry and bio-energy.
2007.9	Interim Measures on Price Subsidy for Biomass Energy and Bio-chemical Material Base	Ministry of Finance	Stipulates the subsidy for forestry raw material base is CNY200/ acre and the subsidy on agricultural raw material base is CNY180 / acre in principle.
2007.6	Opinions on Promoting the Financing Work of Biomass industry in China	NDRC, CDB(China Development Bank)	Builds various innovative financing mode and platform for the development of biomass industry. Establishes and improves the credit guarantee system. Sets up special fund for the development of bio-industry.
2007.1	Interim Measures on Adjustment for Additional Income of Renewable Energy Price	NDRC	Clears the allocation method of subsidy that above the benchmark price, which convert to CNY0.25 /KWh for the on-grid electricity. Power generation projects from straw direct combustion involved in the subsidy scope can simultaneously take the temporary price subsidy when encounter deficit. The subsidy will be CNY0.1/KWh.
2007.3	Enterprise Income Tax Law of the PRC	NPC	Regulates power generation projects which take comprehensive utilization of resources can have income tax reduction of 10%.
2008.12	Notice on Value-added Tax Policies of Free and Comprehensive Utilization and Other Products	Ministry of Finance	Allows power generation from direct combustion of straw projects to get refund of value-added tax.

Data sources: collected and edited by the author.

6. Strategic Planning and Market Potential of the Development of Biomass Energy Industry in China

At the end of 11th FYP period, the government proposed a planning for the development of biomass. Especially during the 12th Five-Year period, National Energy Committee not only formulated *the 12th Five-Year Development Planning on Renewable Energy Resources*, but also proposed special planning for hydropower, wind power, solar energy and biomass energy. In addition, *the Plan on Agriculture Biomass Industry Development (2007-2015)* promulgated by the Ministry of Agriculture and *the National Plan for Forestry Biomass Development (2011-2020)* formulated by the State Forestry Administration, also established strategic goals and key tasks for further development.

6.1. Strategic Goals on Development of Biomass Energy Industry in China

The development planning of biomass energy industry in China includes overall goals on industry development, development of biomass energy material, and strategic goals on technology development. The overall goals are confirmed in *the 12th Five-Year Development Planning on Renewable Energy Resources, the 12th Five-Year Development Planning on Biomass Energy* and *the Plan on Agriculture Biomass Industry Development (2007-2015).* The strategic goals on development of biomass energy material biomass are confirmed in *the Plan on Agriculture Biomass Industry Development (2007-2015)*, and *National Plan for Forestry Biomass Development (2011-2020).* The strategic goals on technology development are confirmed in *the Special Plan on Development of Biomass Energy Technology during the 12th FYP.*

Table 4-13. Strategic goals of biomass development during the 12th Five-Year period

Fields	Scale		Annual production of energy		Standard coal equivalent
	Amount	Unit	Amount	Unit	MT/Year
1. Biomass Generation	130	MW	78	TWh	24.3
Power generation from agricultural/forestry biomass	80	MW	48	TWh	15
Power generation from biogas	20	MW	12	TWh	3.7
Power generation from garbage	30	MW	18	TWh	5.6
2、Bio-gas supply			22 billion	m^3	17.5
Household biogas	50 million		19 billion	m^3	15
Large agricultural residue gas	6,000		2.5 billion	m^3	2
Industrial organic wastewater and sewage treatment plant sludge biogas	1,000		0.5 billion	m^3	0.5
3、BMF	10	MT			5
4、liquid Biofuel					5
Biofuels Ethanol	4	MT			3.5
Bio-diesel and aviation fuel	1	MT			1.5
Total					51.8

The overall goals proposed by *the 12th Five-Year Development Planning on Renewable Energy Resources* remarkably increases the proportion of renewable energy in energy consumption, and proposes that the annual consumption of all renewable energy will reach 478 million tons of sce in 2015 [14]. According to *the 12th Five-Year Development Planning on Biomass Energy*, annual usage of biomass energy will exceed 50 million tons of sce by 2015. Among them, the total installation of biomass power generation will be 13GW with about 78 TWh of total generation, and annual production of biogas, BMF, and bio-liquid fuel will reach 22 billion m^3, 10 million tons and 5 million tons respectively (table 13) [2, 16].

The Plan on Agriculture Biomass Industry Development (2007-2015) put up special plans which focus on the development of biomass energy in rural places and agriculture bio-energy, propose the strategic goals on the development of biomass energy in China. (table 14) [1]

Table 4-14. Strategic goals on biomass development in rural areas and agriculture bio-energy

Fields		2010	2015
Rural biogas	The total amount of rural household biogas projects	40 Million (18 Million newly added)	About 60 Million
	Annual Production of biogas	15.5 Billion m^3	23.3 Billion m^3
Straw energy utilization	Straw molded fuel demonstration site	About 400	
	Annual consumption of straw molded fuel	About 1 Mt	About 20 Mt
	Straw gasification and centralized supply station	About 1,000	About 2,000
	Annual Production of Straw gas	365 Million m^3	730 Million m^3
Energy plants	Moderately develop of the energy plants, and meet the needs of the raw materials for liquid biofuel in China.		

The 12th Five-Year Development Planning on Renewable Energy Resources and the 12th Five-Year Development Planning on Biomass Energy set the strategic goals of the development of biomass energy industry. In addition, National Plan for Forestry Biomass Development (2011-2020) and the Plan on Agriculture Biomass Industry Development (2007-2015) set plans on the development of bio-energy raw materials supply. Especially, the National Plan for Forestry Biomass Development (2011-2020) proposes clear strategic goals (table 15) [3]

Besides, *the National Plan for Forestry Biomass Development (2011-2020)* also proposes that 29 energy forest cultivation demonstration bases will be established by 2015 with a total acreage of 900,000 hectares, among which, 15 are oil energy forest bases with the acreage of 380,000 hectares, 11 are wood energy forest bases with the acreage of 400,000 hectares, and 3 are starch energy f bases with the acreage of 120,000 hectares.

The strategic goals on technology development of biomass energy include: cultivating a number of new efficient varieties of biomass, breaking through core technologies in efficient conversion and utilization of biomass, creating new products of biomass energy, chemicals and materials, acquiring a batch of new patents with independent intellectual property, forming a number of standards for the development of rural renewable energy industry, forming diversified co-production technology system and a group of complete equipment, constructing a number of demonstration bases which reflect the technical characteristics, regional characteristics and product characteristics, cultivating a number of biomass energy enterprises with market competitiveness and a group of talent team contributed to development and industrialization of biomass energy industry, and building sophisticated technology system for the comprehensive utilization of biomass resources [15].

Table 4-15. The overall goals on national forestry development

Fields	2015				2020			
	Scale	Available scale	Biomass Supplied	Standard Coal Equivalent	Scale	Available scale	Biomass Supplied	Standard Coal Equivalent
Oil Energy Forests	2.12 million hectares	0.5 million hectares		700Kt	4.22 million hectares			5.8Mt
Wood Energy Forests	5.24 million hectares	400 million hectares	24Mt	12Mt	9.43 million hectares	7.5 million hectares	45Mt	22Mt
Starch energy forests	1.02 million hectares	0.4 million hectares		90Kt	3.13 million hectares	1.5 million hectares		3Mt

6.2. Key Tasks for the Development of Biomass Energy Industry in China

According to the comparison between the achievements during the 11th Five-Year period and the goals proposed for 12th Five-Year development, we can summarize the key tasks for the development of the biomass energy industry in China. It is evident that biomass power generation, household biogas, BMF and fuel ethanol are all key areas in the coming five years.

The total installation capacity for biomass power will increase to 13 GW from 5.5 GW in 2010, up by 160%. The key tasks include the development of power generation from agricultural/forestry biomass, heat and power cogeneration (CHP), key equipment for power generation from garbage and clean combustion, and power generation from biogas.

The total amount of household biogas would add up to 50 million m^3 in 2015 compared with 40 million m^3 in 2010. The planning puts forward steady promotion for the construction of household biogas. The key tasks include establishing ecological agriculture pattern and improving the comprehensive efficiency of household biogas.

The production of biomass molded fuel (BMF) will increase to 10 million tons in 2015 from 3 million tons in 2010. During the 12th Five-Year Period, the key tasks include promoting centralized heating supply from BMF in northern area and expanding the scale of BMF market.

The production of biofuel ethanol will increase to 4 million tons in 2015 from 1.8 million tons in 2010, up by 2.2 million tons. The key tasks include break-through in key equipments and integration techniques, and the scale economy of cellulosic ethanol by reducing the production cost. Meanwhile, a number of industrialization cellulosic ethanol demonstration projects should be built and form large-scale cellulosic ethanol production bases.

6.3. The Market Prospect of the China's Biomass Industry Development

To achieve these objectives discussed above, besides the policy support from government and regulation guidance, the investment from users, developers, manufacturers and other stakeholders is essential. Based on estimation from *the 12th Five-Year Development Planning on Biomass Energy*, by the end of 12th Five-Year Plan period, the investment of biomass energy industry will reach CNY140 billion. In terms of power generation projects, preferential tariff support will continually be provided [2]. According the specific planning data, we estimate the investment demand in the field of biomass energy from 12th Five-Year Development Planning on Biomass Energy. (table 16)

Table 4-16. Investment demand in the functional field of biomass energy

field	Unit Investment (CNY)	Scale	Investment Required (CNY million)
Fuel ethanol	3500/ton	2,200Mt	7700
Bio-diesel	9000/ton	0.5Mt	4500
Household biogas	3000/household	10 Million	30
Biomass power generation	2810/KW	7500 MW	21075

Note:

1. 1US$=6.3 RMB in 2012.
2. ADB (Asian Development Bank) issued *Rural Biomass Energy 2020 in the People's Republic of China*, which reported the average investment cost of Chinese fuel ethanol is CNY3,500/ton, average investment of bio-diesel is CNY 9,000 /ton, and average household biogas investment is CNY3,000/household.
3. There is no authoritative average investment data for biomass power generation. The average investment is estimated by the author based on two lately biomass power generation cases in China.

7. Barriers and Policy Options for the Development of Biomass Energy Industry in China

The issue of raw material collection is the most important factor to restrict the development of biomass industry in China. There are abundant biomass resources in China, but it is difficult to collect them for utilization purpose. One reason is the scattered distribution of biomass resources in China, due to household contract responsibility system, a decentralized farming mode. In turn this system results in the scattered distribution of straw and stalks, which is the majority of the biomass resources. On the other hand, because the biomass electricity price is not high enough, the biomass resources price is not attractive to the farmers. Take maize stalks as an example, power plants typically pay CNY200-260/ton, while the purchase price is only about CNY100/ton by collection station from farmers. A farmer can obtain 3 tons of maize stalks from 8 acres of farmland. So he can get CNY300 from the collection station, which is not profitable enough if considering the oil cost and transportation expense. In this situation he would rather choose to burn up the maize stalks to get fertilizer [16]. Therefore, a well-designed incentive system and efficient utilization of biomass raw materials are essential to overcome resource obstacles in China. On the one hand, the policy implementation should be enhanced to prevent crop and straw incineration, which causes not only the loss of biomass energy sources, but also environment damage. On the other hand, the purchase price of the raw material resources should be raised, but it is basically related to market price of biomass products. If the subsidies to the biomass products are not high enough, a feasible way to is provide subsidies to crop straws providers (peasants).

In terms of medium and small sized biomass energy enterprises, the policy requirement to be eligible to get subsidy is too much. For example, according to provisions issued by the Chinese Ministry of Finance, an enterprise can get subsidies of CNY140 /ton if the registered capital of the enterprise is more than CNY10 million and the consumption of crop straw reaches ten thousand tons per year. This term is no doubt too demanding for some medium and small sized biomass energy enterprises. Meanwhile, the government sets a grid-connection capacity of 500KW, which is much higher than existing installed capacity of 100KW. In addition, due to the complicated implementation process, local grid companies are unwilling to have small power plants connect to the power grid. Therefore, it is necessary to lower the eligible qualification of subsidies and grid integration to ensure that the medium and small biomass enterprises can get opportunities to operate in this industry.

Emphasis on construction subsidies but neglect of maintenance and operation supervision results in lack of technical equipment investment for many biomass projects. The subsidy policy for biomass projects in China focuses on project investment, largely ignoring maintenance and operation supervision. The consequence is that some enterprises acquire subsidies by building biomass projects but leave the assets unused, which cause a waste of supporting fund. In addition, because the government supervision is not enough, enterprises are reluctant to assume the high costs for R&D and introduce advanced equipments, which result in unstable operation and low efficiency during the production process. Thus, the subsidy policy and management mode should be changed, shifting the focus from project grants to subsidize biomass products or generation, and attracting capital investment through raising the price of biomass products. From the management perspective, the emphasis on project approval should extend to the management and maintenance during the whole process of the project.

Existing tariff for biomass power generation is benchmarked based on desulfurization coal-fired units in the provinces with additional subsidy price of CNY0.25/KWh. In July of 2010, the NDRC issued a Policy on improving agriculture and forestry biomass power price, specifying that for the new agricultural/forestry biomass power generation projects without bidding investor, the benchmark price is set as CNY0.75/KWh (including tax). According to the study by the Asian Development Bank, most of China's biomass power plants under construction can not be cost-effective even with the government subsidy [13]. Therefore, to sustain biomass generation, the government should switch the price policy that can provide enough incentive to the investors.

Current management system can't effectively meet the management and services requirements of biomass projects. In China, the NDRC is responsible for project approval and planning and the local government is responsible for project organization and implementation. For biomass energy projects in rural areas, the Ministry of Agriculture at the central level is responsible and complete management and service system is in place. All provinces, autonomous regions and municipalities as well as more than 90 percent of the counties have rural energy management departments, responsible for supervising the construction of biogas digesters, organizing technical training and promoting new technologies. It is estimated that there are a total of more than 50000 staff in these departments [13]. However, in liquid biofuel and biomass power generation fields, the management and service system is not complete. We recommend strengthening the coordination among government departments and forming complete management and service system. The system should ensure the connection of government between upper and lower levels and the coordination among different government departments.

CONCLUSION

China has made considerable progress in biomass energy development, especially in aspects of rural biogas, biomass power generation and fuel ethanol. But take consideration of the abundant biomass resources and the environment and energy pressures in China, there is still tremendous space for future improvement. Chinese government has recognized the issues and made ambitious strategic goals for biomass energy industry during the 12th Five-Year Plan period. But there are still some institutional obstacles, including how to effectively manage biomass materials and stimulate farmers; how to provide subsidy incentives for biomass product price; how to manage the projects in a comprehensive way; and how to establish management and service system, etc. Corresponding countermeasures are proposed in this chapter to solve these issues.

REFERENCES

[1] Ministry of Agriculture. The Plan on Agriculture Biomass Industry Development (2007-2015)[online]. 2007[cited 2007 6 18]. Available from: http://www.moa.gov.cn /zwllm/ghjh/201006/t20100606_1533134.htm.

[2] National Energy Administration. The 12th Five-Year Development Planning on Biomass Energy[online]. 2012[cited 2012 12 28]. Available from: http://www.gov.cn /zwgk/2012-12/28/content_2301176.htm.

[3] State Forestry Bureau. National Forestry Biomass Development Plan (2011-2020)[online]. 2013[cited 2013 5 28]. Available from: http://www.ccchina.gov.cn /archiver/ccchinacn/UpFile/Files/Default/20130614090957368659.pdf.

[4] NPC(National People's Congress). The outline of the 11th Five-Year Plan for National Economic and Social Development[online]. 2006[cited 2006 3 14]. Available from: http://www.gov.cn/gongbao/content/2006/content_268766.htm.

[5] NDRC(The national development and reform commission). Long-and Medium-Term Development Planning on Renewable Energy Resources[online]. 2007[cited 2007 8 31]. Available form: http://www.sdpc.gov.cn/zcfb/zcfbtz/2007tongzhi/W020070904607346044110.pdf.

[6] CAS(Chinese Academy of Sciences)The current situation of the development of bioenergy and future technology[online]. 2010[cited 2010 5] Available from: http://www.qibebt.cas.cn/xscbw/xxcp/ztbg/201110/P020111026499140612013.pdf.

[7] NDRC; Ministry of Agriculture; ministry of finance. the notice on the implementation scheme of comprehensive utilization of crop straw during the 12th five-year-plan[online]. 2011[cited 2011 11 29]. Available from: http://www.mof.gov.cn/zhengwuxinxi/zhengcefabu/201112/t20111221_617842.htm.

[8] NBS (National Bureau of Statistics). China Statistical Yearbook 2012[online]. 2012[cited 2012 9 26]. Available from: http://www.stats.gov.cn/tjsj /ndsj/2012/indexch.htm.

[9] Hu R. *the technology roadmap of biomass energy in China.* Edition:1. Beijing: China Environmental Science Press; 2011.

[10] Glinfo (website). The development status and prospects of fuel ethanol in China[online]. 2012[sited 2012 11 20]. Available from: http://finance.glinfo.com/12/1120/16/3CFF17363D1428B1.html.

[11] Yan Zh. Biodiesel development obstacles need to be solved[online]. 2012[sited 2012-12 31]. Available from: http://www.39kf.com/yyjj/biotechnology/03/2012-12-31-852068.shtml.

[12] China Power (website). The 12th Five-Year Development Planning on Biomass Energy issued, total installed capacity will reach 13million KW by 2015[online]. 2013[cited 2013 1 4]. Available from: http://www.chinapower.com.cn/article/1223/art1223876.asp.

[13] Asian Development Bank. Rural Biomass Energy 2020 in the People's Republic of China [online]. 2010[cited 2010 6]. Available from: http://www.adb.org/publications/2020.

[14] National Energy Administration. The 12th Five-Year Development Planning on Renewable Energy Resources[online]. 2013[cited 2013 2 1]. Available from: http://www.ce.cn/cysc/ny/zcjd/201302/01/t20130201_21331890.shtml.

[15] Ministry of Science and Technology. The 12th five-year-plan key special biomass energy technology development plan [online]. 2012[cited 2012 7 3]. Available from: http://www.biogas.cn:8080/Z_Show.asp?ArticleID=2194.

[16] Gao F. why development of power generation from biomass struggling? *Liaoning Urban Rural Environmental Science & Technology*, 2013 33(10), 32-33.

In: Renewable Energy in China
Editors: J. Yuan, G. Luo and J. Chen
ISBN: 978-1-63117-610-4

Chapter 5

DEVELOPMENT AND UTILIZATION OF MARINE RENEWABLE ENERGY RESOURCES IN CHINA

Guoliang Luo*
School of Economics and Management,
North China Electric Power University, People's Republic of China

ABSTRACT

China is abundant in marine renewable energy resources, which, however, are developed insufficiently. At present, the development of marine wind energy is generally mature as commercial operation is realized. Only experimental facilities are available for the utilization of tidal energy, wave energy, tidal current energy and temperature difference energy, and only principle experiments are conducted for salt difference energy. China has made great headway in developing and using marine renewable energy resources, which is owed mainly to the Chinese government that has offered the field great support in legislation, planning, institution and funding. In 2010, the Ministry of Finance set up a special fund for marine renewable energy, annually investing RMB200 million in the demonstration projects of power generation systems as well as key technologies and experiments. This greatly advances the researches on marine technologies in China. However, China still lags far behind developed countries in most marine renewable energy technologies, and shall take effective steps in technologies, funds, policies and management to narrow the gap and accelerate development of marine renewable energy.

Keywords: Marine renewable energy resources, technologies, policies planning

1. INTRODUCTION

China began to develop marine renewable energy technologies in the 1960s, and has made great headway in developing and using marine renewable energy resources after five

* Email: lgl365365@126.com.

decades of efforts. China is abundant in marine renewable energy resources, which however are insufficiently developed. At present, development of marine wind energy is basically mature as commercial operation is realized. Only experimental facilities are available for the utilization of tidal energy, wave energy, tidal current energy and temperature difference energy, and only principle experiments are conducted for salt difference energy.

Marine renewable energy resources are important to China. So, developing and using such resources can promote the development of coastal and marine economy, ease the pressure in energy supply, reduce environmental pollution and improve the energy structure. Island residents in China face a short supply of energy, which concerns the development of rural economy in the coastal areas and islands. Therefore, developing marine renewable energy is a practicable way to solve the energy problem facing them. Furthermore, developing marine renewable energy can help China strengthen its management and control over the South China Sea as this requires development and protection of the islands.

This chapter aims to analyze the current development and utilization of marine renewable energy in China as well as the policy environment, obstacles and countermeasures. The paper includes six parts. Section 2 introduces the current development and utilization of marine renewable energy at home; Section 3 reviews domestic policy environment for marine renewable energy; Section 4 analyzes the bottlenecks facing the development and utilization of marine renewable energy at home; Section 5 proposes the countermeasures for developing marine renewable energy; Section 6 offers conclusions.

2. Current Development and Utilization of Marine Renewable Energy at Home

Marine renewable energy is an important clean energy hidden in the ocean. It comes in different forms, including tidal energy, wave energy, tidal current energy, temperature difference energy, salt difference energy and offshore wind energy.

2.1. Marine Renewable Energy Reserve and Distribution in China

China has a continental coastline of more than 18,000 km and an island coastline of 14,000 km. With 6,500 islands of different sizes, it has over 4.7 million sq km of sea areas and is abundant in marine renewable energy that is up to 500 million kW [1].

2.1.1. Tidal energy

Tidal phenomenon refers to the periodic movement of seawater caused by the tide generating force of celestial bodies (mainly the moon and the sun). We commonly call the vertical rise and fall of the sea level as tide, and the horizontal movement of seawater as tidal current.

It is estimated that the theoretical reserve of tidal energy in China is 190 million kW, 20.98 million kW of which can be developed. Most of the tidal energy is concentrated in the coastal areas of East China Sea, particularly Zhejiang and Fujian Provinces, which makes up 88% of the national total. A total of 424 dam sites with the installed capacity above 200 kW can be developed; In particular, tidal power stations with the installed capacity reaching 100,000 kW or 1 million kW can be built in Yueqing Bay, Xiangshan Bay, north branch of the Yangtze River estuary and the estuary of Qiantang River (see table 5-1) [2].

Table 5-1. Tidal energy in China's coastal provinces, municipalities and autonomous regions

Provinces, municipalities and autonomous regions	Theoretical reserve		Developed reserve	
	Installed capacity (MW)	Annual power generation (MW)	Installed capacity (MW)	Annual power generation (10^4 kW)
Liaoning	1936	577	586.2	16.24
Hebei	6.4	1.3	4.7	0.09
Shandong	408	121.9	117.8	3.63
Jiangsu	----	----	0.8	0.04
Shanghai	2946	879.3	704	22.80
Zhejiang	289683	8626.8	8801.6	264.04
Fujian	33290.2	9936.6	10324	283.32x
Guangdong	----	467	648.8	17.20
Guangxi	----	465.9	387.3	10.92
Total	----	21075.8	21575.2	618.68

2.1.2. Wave Energy

Waves are shaped by wind, air pressure and water gravity, and have certain kinetic energy and potential energy. In China, the coastal wave energy features uneven power density.

About 130 million kW of wave energy can be potentially developed. As the central part of Zhejiang, Taiwan Strait, the areas north of the Haitan Island, Fujian, Bohai Strait and Xisha Islands boast the largest wave energy reserve, it's practicable to build wave energy-based power stations in Zhejiang, Fujian, Hainan and Taiwan (see table 5-2) [2].

In China, Taiwan has the richest offshore wave energy that is up to 4.29 million kW and accounts for one third of the national total.

It is followed by Zhejiang, Guangdong, Fujian and Shandong, which has the wave energy between 1.61 million and 2.05 million kW (7.06 million kW in total, representing 55% of the national total). Other parts of the country see less wave energy, which is between 144,000 and 563,000 kW [3].

2.1.3. Tidal Current Energy

Tidal currents refer to the relatively stable movement of seawater in undersea tunnels and straits as well as the regular seawater movement caused by tide.

The coastal areas in Zhejiang, Fujian and Shandong are among the areas richest in tidal current energy in the world.

In particular, the tidal current velocity in some sea areas of Zhoushan Islands is 2-4 m/second, with the energy flux density equivalent to 20-40 m/second (9-12 levels or above). So, the development potential is huge (see table 5-3) [2].

Table 5-2. Distribution of wave energy in part of China

	Taiwan	Zhejiang	Guangdong	Fujian	Shandong	Hainan	Jiangsu	Liaoning	Other provinces
Resources reserve (MW)	4291	2053	1740	1660	1610	563	291	255	381
Proportion (%)	33.4	16	13.5	12.9	12.5	4.4	2.3	1.2	2.9

Table 5-3. Distribution of tidal current energy in part of China

	Taiwan	Zhejiang	Guangdong	Fujian	Shandong	Hainan	Hainan	Liaoning	Other provinces
Resources reserve(MW)	2283	7094	360	1280	1195	276	276	1140	381
Proportion (%)	18.2	56.6	2.8	10.2	9.5	2.2	2.2	9.1	2.9

2.1.4. Temperature Difference Energy

The temperature difference energy in seawater refers to the energy arising from the temperature difference between the surface seawater and the deep sea water. The temperature difference energy that can be developed is mainly distributed in the east side of the East China Sea and the South China Sea. The black tide area in the east side of the East China Sea is more than 1,000 m deep, and has the temperature difference of 20°C or above because of the passage of warm black tide. It is estimated that the theoretical reserve of the temperature difference energy is about (1.08～1.75) ×1018 KJ, and the reserve that can be used technologically is (54～112) ×1015 KJ. The theoretical reserve of the temperature difference energy in South China Sea is (13.0～13.3) ×1018 KJ, and the reserve that can be used technologically is (1.04～1.06) ×1018 KJ [4].

2.1.5. Salt Difference Energy

China's offshore salt difference energy resources are mainly distributed in the Yangtze River and the areas south of the Yangtze River as well as the coastal areas where there are estuaries of major rivers. The theoretical reserve is about 3.58x1015 KJ, and the theoretical power 114 million kW. In particular, the combined salt difference energy of the Yangtze River estuary and the Pearl River estuary accounts for 81.24% of the national total [5].

2.1.6. Offshore Wind Energy

China's offshore wind energy is three times that of onshore wind energy, and the wind energy that can be developed and used totals 750 million kW. The southeastern coastal areas and islands from the Yangtze River to the Nan'ao Island are the largest wind energy area at home. Areas rich in wind energy include Shandong, Liaodong Peninsula, the coastal areas of the Yellow Sea, South China Sea west of the Nan'ao Island, Hainan Island and South China Sea islands.

2.2. Development of Marine Renewable Energy in China and Relevant Researches

China has accelerated development of marine renewable energy in recent years. Construction of Changdao offshore wind farm in Shandong and Phase I of Rudong offshore demonstration wind farm in Jiangsu has been commenced; Construction of Donghai Bridge offshore wind farm in Shanghai has been successfully completed; Zhejiang Sanmen tidal power station and Fujian Bachimen tidal power station have been officially launched; A marine microalgae biological energy project has settled down in Longgang, Shenzhen. All these events demonstrate the significant progress China has made in developing marine renewable energy [6].

2.2.1. Tidal Energy

Among all the technologies related to marine renewable energy, China has the maturest technologies in tidal energy. The high cost of building tidal power stations leads to high power generation cost. In 2012, there were two tidal power stations still in operation in Zhejiang, i.e., Jiangxia and Haishan tidal power stations, which have a combined installed capacity of 4,150 kW. Jiangxia tidal experimental power station has the biggest size, with the installed capacity as 3,900 kW, and represents the highest level of China's tidal power generation technologies. China has independently designed and manufactured the tidal power generator units with the single installed capacity reaching 26,000 kW, which can resist against tough marine environment and are basically commercialized. By the end of 2011, tidal power totaled about 172 million kWh on a cumulative basis [7].

2.2.2. Wave Energy

Despite the late start, China has made great headway in the research and utilization of wave energy. After three decades of efforts, China has built 100 kW oscillating water column and 30 kW pendulum wave power stations for the experimental purpose. The shoreline wave power generation technologies have taken the lead in the world, and the micro-type wave power generation devices used in beacon lights have been commercialized [8]. Supported by the 11th Five-year Plan for the development of science and technologies, China has initiated construction of two experimental power stations each with an installed capacity totaling 100 kW or above. In April 2010, China's first wave power generation project settled down in Gezhou Island, Guangdong Province, with the first generator officially put into use in early May. About RMB350 million of funds would be invested to build a 50,000 kW plant that can generate 300 million kW·h of power annually. This is the first wave power station China has ever built with foreign technologies [7].

2.2.3. Tidal Current Energy

China developed a 70 kW floating experimental device and a 40 kW bottom supported device for vertical axis tidal current power stations respectively in the 8th Five-year Plan period and the 9th Five-year Plan period; During the 11th Five-year Plan period, it launched construction of 100 kW vertical axis and horizontal axis tidal current energy power stations and corresponding experimental work. By far, China has made great progress in the transformation of tidal current energy, design of power generation systems, analysis of performances, R&D of key technologies and experimental devices as well as the fixing of underwater devices.

Technologies related to tidal current energy haven't been mature, and the industry hasn't taken shape. At present, no tidal current energy projects have seen a capacity of more than 100 KW, and no mature products have been developed for tidal current power generation. Moreover, the power generation cost is high, and most projects are subsidized by the State rather than funded by enterprises. But tidal current power generation enjoys bright prospects in the islands rich in tidal current energy, and China still has huge room for technological improvement in this regard.

2.2.4. Temperature Difference Energy

China is still at the infant stage in the technologies related to temperature difference energy. No experimental power stations have been built under actual sea conditions. But once mature, the technologies will play a big role in the development of Xisha and Nansha Islands. In the early 1980s, China started researches on temperature difference power generation in Guangzhou, Qingdao and Tianjin. In 1986, Guangzhou reported completion of the open-type experimental analog device for the transformation of temperature difference energy that enables power transformation. In 1989, the experimental research on fog drop improvement and circulation was completed. In summary, China's development and utilization of temperature difference energy is still at a stage of R&D [9].

2.2.5. Salt Difference Energy

China is still at an initial stage in the research on salt difference energy, and the industry hasn't come into being. It is estimated that the salt difference energy at all the estuaries throughout the world totals 30 TW, 2.6 TW of which can be actually utilized. In China, the salt difference energy is about 1.1×108 kW, which is mainly distributed at the estuaries of major rivers. Furthermore, many inland salt lakes in Qinghai and some other areas can also be utilized. The US and Israel take the lead in the researches on the salt difference energy, and China, Sweden and Japan also carried out some researches. But on the whole, such researches are still in labs, and no experiments about energy transformation have been conducted yet.

2.2.6. Offshore Wind Energy

China's application of wind energy technologies is quickly shifting from mainland areas to coastal areas and shoals, and further to offshore deepwater areas. Once the breakthrough in underwater installation and platforms is achieved, offshore wind energy technologies will swiftly be applied to offshore areas. Construction and trial use of wind energy installation ships as special ships have been completed. In the near future, China will focus on the coastal areas of Jiangsu for offshore wind energy development.

In July 2010, all the generators in Donghai Bridge offshore wind farm were put in place and connected to the grid. This is the first large offshore wind farm in both China and Asia, with the installed capacity as 102,000 kW, the annual on-grid power 267 million kWh and the project investment RMB2.37 billion. Meanwhile, the 3.6 MW large offshore wind power generator units China independently developed was put into use at the electrical heavy equipment port base in Shanghai, with the wind turbine having a diameter of 116 m and the single generator generating 9 million kWh of power annually [5]. Besides, Shanghai Electric has set up the wind turbine test center at the port manufacturing base, which, provided with full power test beds, has the only port experimental wind farm among all the main engine plants at home and can complete the most comprehensive inspections and tests related to wind turbines.

By the end of 2012, five offshore wind power projects were built at home, having an installed capacity of 298,000 kW in total; Sixteen projects with an installed capacity of 3.4 million kW were approved by the National Development and Reform Committee, and the pre-feasibility study reports of 24 projects with a combined installed capacity of 5.65 kW were completed or underway. The said 45 projects have a total installed capacity of 9.35 million kW [10].

2.3. Comparison between China and Developed Countries in the Technologies Related to Marine Renewable Energy-Based Power Generation

China's technologies in developing marine energy, except the tidal energy, mostly lag behind the world's advanced levels. At present, researches on wave energy and tidal current energy are still at the initial stage of demonstration, almost 3-6 years behind the world's advanced level. Researches on temperature difference energy are only conducted in the Institute of Oceanology, while the researches on salt difference energy almost come to a halt. In 2010, driven by the State Oceanic Administration, the Ministry of Finance set up a special fund for marine renewable energy and annually inputs RMB200 million. This supported six independent power grid demonstration projects, three on-grid power demonstration projects, 39 key technology researches and experiments, two industrialization demonstration projects and construction of 14 standard and service systems related to marine renewable energy in 2010-2011 [10]. This greatly advances researches on marine technologies as well as development and utilization of marine energy. But China still lags behind developed countries in most technologies related to marine renewable energy-based power generation [11].

(1) China has the technologies to develop large tidal power stations. The generator units at Jiangxia power station in China were designed based on Rance tidal power station in France and the actual conditions of Jiangxia, with the single generator capacity as 500-700 kW. The general technological level is on a par with that of Rance power station. Through constant R&D and practices, China has been geared to the world's advanced level in the design and manufacturing of tidal power generator units, and presently has the capability to manufacture tidal power generator units with the installed capacity reaching 10,000 kW or above.

(2) China doesn't lag far behind developed countries in the R&D of technologies related to power generation with tidal current energy, as both are at the experimental stage. In 2009, the commercialized tidal current power generation device SeaGen developed in the UK was put into normal operation, and the power generated based on the technology will total 500,000 kW by 2015; In the South Korea, a 300,000 kW tidal current power station based on the said technology is under construction [7]. In China, the demonstration size is relatively small at present. Following the research on the 70 kW tidal current power generation facility by Harbin Engineering University, a 220 kW tidal current power generation facility is under construction now. This shows that China lags far behind developed countries in the sizes of tidal current power generator units and the installed capacities.

(3) Wave power generators are still under development in China, while have been scaled up in developed countries. Domestic research on the technologies for power generation with wave energy dates back in the 1970s, and has made headway since the 1980s. The micro-type wave power generators used in beacon lights have tended to be commercialized. Due to the small batches, short working life and poor reliability, wave power generation involves high cost. In 2009, the 750 kW offshore wave energy converters were put into operation in the UK, with the planned installed capacity as 2,250 kW; Forty-eight wave power stations are being constructed or planned in the US, annually generating 2.1 billion kWh of power [7]. The said technologies represent the trends for the development of marine energy. During the 11th Five-year Plan period, China organized researches and demonstration of key technologies respectively for 100 k floating wave power stations, 20 kW ocean current energy devices and 150 kW tidal current energy power stations, which accumulated technologies for the development of wave energy and tidal current energy [7].

(4) The research and development of temperature difference energy in China still lag far behind the world's advanced level. The US, Japan, Netherlands, France, UK and India have designed or planned to build temperature difference energy-based power stations with the installed capacity totaling 10,000 kW or 100,000 kW. A 5,000 kW ocean temperature

difference energy-based power station that integrates power generation, aquaculture and entertainment has been built in the coastal areas in the eastern Taiwan. Chinese mainland has only seen completion of principle experiments.

(5) China has a later start in generating power with marine wind energy when compared with foreign countries. China didn't have its first offshore wind farm till 2007, while the world's first offshore wind farm was constructed in 1991 by Denmark. The IPRs of the key technologies for ocean wind energy-based power generation mostly are not owned by China. MW-level wind turbines and key parts and components are mainly imported or manufactured on a franchise basis, and motor control technologies are mainly imported, too.

3. Policy Environment for the Development If Marine Renewable Energy in China

China needs to develop marine renewable energy, with the aim to meet the climate change, realize energy conservation and emission reduction, foster strategic emerging industries and accelerate the transformation of economic growth modes. Over the past years, China has made progress in developing marine renewable energy, which is attributed to the sound policy environment, which includes legislation, planning, management institutions and funding.

3.1. Legislation

The Renewable Energy Law revised in 2009 explicitly included marine energy as part of the renewable energy. The Central Government also developed a package of regulations [7], including the Interim Measures for Administration of Renewable Energy-generated Power Tariff and Cost Distribution, the Interim Measures for Administration of Special Funds for the Development of Renewable Energy, the Regulatory Measures for Grid Enterprises' Full Purchase of Renewable Energy Electricity, the Interim Measures for Allocation of Income from Surcharges on Renewable Energy Power Prices and the Interim Measures for Management of the Special Funds for Renewable Energy Development.

3.2. Management and Planning

As the competent authority governing the ocean of China, the State Oceanic Administration assumes researches, application and management of marine renewable energy. In 2010, it set up the Center for Administration of the Development and Utilization of Marine Renewable Energy, which is responsible for administering the technologies related to marine renewable energy.

Since 2010, the Chinese government has formulated a series of development plans [7], including the 12th Five-year Plan for the Development of Renewable Energy (2011-2015), the 12th Five-year Plan for the Development of China's Marine Affairs (2011-2015) and the 12th Five-year Plan for the Development of China's Marine Economy (2011-2015). The 12th

Five-year Plan for the Development of Renewable Energy (2011-2015) defined the goals for the development of marine renewable energy for the first time, and made "accelerating the advance in marine renewable energy technologies" as one of the key tasks in the plan. The concrete tasks include:

- Improve technologies to develop and use marine renewable energy, and gradually scale up use of marine renewable energy.
- Construct independent marine renewable energy-based power stations on islands that complement wind energy and solar energy;
- Build 1-2 tidal power stations with an installed capacity of 10,000 kW or above and numerous on-grid tidal current power stations for the demonstration purpose.
- Built power stations based on various kinds of marine renewable energy with the installed capacity totaling 50,000 kW by 2015.
- Gradually develop the capabilities to develop the technologies and equipment related to marine renewable energy.

3.3. Financial Support

Since the 10th Five-year Plan period (2001-2005), the Chinese government has given great support to the development of wave energy and tidal energy and subsidized key technologies and projects (See table 5-4 and 5-5)[12].

Table 5-4. Main projects for wave technologies funded by S&T program since 2001

Period	Project name	Project source	Assumed by
2001-2005	Vibration overtopping wave power generation system	National Key Technology R&D Program	Guangzhou Institute of Energy Conversion, Chinese Academy of Sciences
2006-2010	Floating duck wave energy-based power generation system	863 Program	Guangzhou Institute of Energy Conversion, Chinese Academy of Sciences
2006-2010	Island renewable energy independent energy system	863 Program	Guangzhou Institute of Energy Conversion, Chinese Academy of Sciences
2006-2010	Daguan Island floating wave power station	National Key Technology R&D Program	National Ocean Technology Center
2006-2010	Duck wave energy demonstration power station	National Key Technology R&D Program	Guangzhou Institute of Energy Conversion, Chinese Academy of Sciences
2006-2010	Floating direct-driven power generation system	863 Program	Guangzhou Institute of Energy Conversion, Chinese Academy of Sciences
2006-2010	Dish-type overtopping wave power generation system	863 Program	Ocean University of China (OUC)

Table 5-5. Main tidal current energy projects supported by the Chinese government since 2001

Period	Project name	Project source	Carried out by
2001-2005	40 kW independent power generation system with ocean bottom tidal current energy	863Program	Harbin Engineering University
2006-2010	Tidal current power station	863 Program	Harbin Engineering University
2006-2010	150 kW tidal current energy-based power generation system	National Key Technology R&D Program	Harbin Engineering University
2001-2005	Undersea instrument power supply technology	863 Program	Northeast Normal University
2006-2010	Key technologies for power generation with surface tidal current energy	863 Program	Northeast Normal University
2006-2010	Technological demonstration of 20 kw tidal current energy-based power generation	National Key Technology R&D Program	Northeast Normal University
2006-2010	Flexible blade tidal current energy-based power generation system	863 Program	Ocean University of China
2006-2010	Research on ocean current power generation based on electrohydraulic control technologies	863 Program	Zhejiang University
2006-2010	Key technologies for ocean current-based power generation and power supply with new energy on islands	863 Program	Zhejiang University

4. Bottlenecks Facing China in the Development and Utilization of Marine Renewable Energy

China has offered sound environment for the development and utilization of marine renewable energy in technologies, funds, policies and management. But it's undeniable that China still faces challenges in technological development, information, industrialization, economic returns and management systems [13].

4.1. Relatively Backward Technologies and R&D

China's input in marine renewable energy has long been insufficient and mature technologies haven't been developed yet. Domestic researches on marine renewable energy date back to the 1960s, and have made great progress after five decades of development. However, the relatively mature technologies are presently confined to the utilization of tidal energy and wind energy. On the whole, China's technologies related to marine renewable energy lag behind the world's advanced levels in terms of efficiency, reliability, economic performance and demonstration sizes [14].

China has a late start in the R&D of marine renewable energy, and the technologies are not advanced and mature as a whole. This is partly attributed to the relatively independent work by the research organizations of marine renewable energy, which are mostly scientific research institutes and schools of higher learning. With little cooperation, they keep their own key technologies confidential and fail to share knowledge. This greatly restricts the development and utilization of marine renewable energy.

4.2. Data Aging in the Development and Utilization of Ocean Resources

China still lacks accurate and detailed data about ocean resources and corresponding engineering environment. In developed countries, surveys of resources and ocean environment are first conducted to offer detailed data for technological R&D and investment decision-making. Most of the data released in China are the survey results around the 1990s. First, such data are technologically backward, less precise and fewer in quantities; Second, the surveys were mainly concentrated in coastal areas, which accounted for only 40% of the offshore areas; Third, they can't accurately reflect basic situations at present due to the significant changes in China's offshore resources and environment conditions over the past two decades [15]. Outdated data about ocean resources and conditions and the lack of authoritative literature that can illustrate the potential power of marine renewable energy become the information obstacles that curb the demonstration and scaling-up of marine renewable energy technologies.

4.3. Low Level Technology Transfer

A survey of global marine renewable energy shows that American and British marine renewable energy companies account for 52% of the world's total, and 94% of the enterprises were newly established; Among all the surveyed companies, 27% of the R&D funds come from governments, while 73% from private sectors [16]. By contrast, China's development of marine renewable energy is mainly funded by the Central Government [17]. Enterprises haven't become developers of marine renewable energy, and R&D is disconnected with application, as marine renewable energy technologies haven't been efficiently industrialized.

4.4. Economic Obstacles

Effectively transferring marine renewable energy into energy that can be used requires high-end technologies as well as lots of funds. According to the analysis by Carbon Trust, completing the transformation from fundamental researches on marine renewable energy to full demonstration and application involves RMB120 million at the least [15]. Compared with other energy, development of marine renewable energy features higher risk, more investment and lower returns in the short term, thus can hardly arouse the enthusiasm in enterprises and private sectors. Therefore, development of marine renewable energy must rely on financial support from the governments. In 2010, driven by the State Oceanic Administration, the Ministry of Finance set up a special fund for marine renewable energy, annually inputting RMB200 million in the researches and demonstration of relevant technologies. But that is far from enough to meet R&D needs [15].

In general, the economic performance of marine renewable energy is poorer than that of conventional energy in the short term. Let's take tidal energy for example. The investment per kW is about RMB20,000-30,000 for a relatively big tidal power station, compared with RMB4,000-5,000 for a relatively big thermal power station. The investment in a tidal power station is 5-6 times that in a thermal power station, 2-3 times that in a wind power station, and 3-4 times that in hydropower station [13]. Therefore, development of marine renewable energy has poorer economic performance from the perspective of investment.

4.5. Policy and Institutional Obstacles

In January 2006, China's Renewable Energy Law officially took effect. Afterwards, China successively launched a policy package including the Long and Medium-term Plan for the Development of Renewable Energy and the 11th Five-year Plan for the Development of Renewable Energy, which not only specified the general goals for China's renewable energy development, but also set long and medium-range goals for some concrete industries. To achieve the goals, China also released the Renewable Energy Law, the Trial Measures for Management of Allocation of Renewable Energy Power Tariff and Expenses and the Interim Measure for Allocation of Additional Revenue from Renewable Energy Power Tariff, which included incentives for renewable energy power generation. There are three main principles:

- Power grid enterprises must purchase the feed-in power generated from renewable energy in full amount;
- Power grid enterprises purchase the power generated from renewable energy based on the feed-in power tariff that combines reasonable cost and reasonable profits;
- The difference between renewable energy power tariff and conventional tariff is allocated in power grid across the country.

However, concrete implementation regulations, such as those about tariffs of power generated with marine renewable energy, are insufficient [18]. In actual operation, many difficulties are challenging the development of marine renewable energy.

Institutionally, there are problems such as management by several authorities and cross-sector supervision. The researches, development and management of marine renewable energy is in the charge of the State Oceanic Administration, project approval and tariff formulation the National Development and Reform Commission, industrial supervision and law enforcement the State Electricity Regulatory Commission, fiscal support and tax preference the Ministry of Finance and the State Administration of Taxation, State asset management and performance assessment the State-owned Assets Supervision and Administration Commission, and grid connection and price settlement involve power grid enterprises. The core power related to the development and management of marine renewable energy is distributed in many sectors, which gives rise to many problems such as overlapping functions and unclear definition of rights and responsibilities. So, competent authorities and their responsibilities shall be made clear as soon as possible, which will help the scientific development of marine renewable energy.

5. Counter Measures for the Development and Utilization of Marine Renewable Energy

(1) Improve the policies and regulations concerning development of marine renewable energy. Foreign experience shows that marine renewable energy can't be successfully developed without the vigorous support from governments. The Renewable Energy Law revised in 2009 specified the policies concerning R&D, industrial development, investment, prices and taxes related to renewable energy, and defined the responsibilities, rights and obligations of governments, enterprises and users in the development and utilization of renewable energy. This offers sound environment for the development of marine renewable energy. To accelerate the development of marine renewable energy in China and support the economic and social development in islands and coastal areas, China shall also put in place and improve the regulations and supporting measures related to marine renewable energy.

(2) Coordination and comprehensive planning. Developing and utilizing marine renewable energy requires coordination among many authorities. Bu China hasn't set up a special authority for the comprehensive development of marine renewable energy. As the power generation involves tariff formulation, cost distribution and some other aspects, development of marine renewable energy can be smoothly implemented only with effective coordination and cooperation among ocean management, commodity prices and finance authorities and power grid enterprises.

The experience of developed countries in developing marine renewable energy fully shows that the marine renewable energy industry shall be guided by scientific and reasonable planning. In China, the planning shall first be subject to the strategic planning for marine development, coordinated with the marine environment and economical and social development in various coastal areas, and based on the features of marine renewable energy. In the near term, China shall conduct resources prospecting based on the general investigation of marine renewable energy, formulate the plan for the development of marine renewable energy, and develop demonstration projects for development of marine renewable energy by launching special funds and making breakthroughs in key technologies; In the medium term, China shall strengthen R&D and construction of pilot projects, scale up mature technologies and realize industrialization; In the long term, China shall make marine renewable energy one of the main clean energy resources and develop into a strong country in the development and utilization of marine renewable energy.

(3) Develop preferential policies to draw social funds. R&D of the technologies related to marine renewable energy involves high risks and great input, and government investment alone is far from enough. So, this requires greater investment from governments and introduction of other funds as well. Successful foreign experience shows that social funds are also needed for joint development of marine renewable energy. While having clear knowledge of resources and engineering conditions, China shall set the goals for long-term development and develop stable preferential policies, such as those regarding preferential

taxes, tariff subsidies and protection of investors' interests, to encourage input by private funds and support the development and utilization of renewable energy.

(4) Beef up support for demonstration projects of technological innovation. Technological innovation is fundamental to maintaining vitality, while project demonstration is key to the scaling-up of marine renewable energy technologies. Demonstration experiments are of great significance to improving the operational efficiency of power stations, lowering power generation cost and making technologies more pragmatic. Foreign experience tells us that the governments of developed countries pay great attention to both the R&D of renewable energy technologies and the demonstration and scaling-up of new technologies. For example, Japan established the technological system for new energy that integrates R&D, demonstration and scaling-up. Through demonstration, scaling-up and industrialization of advanced technologies, China strives to rank among the world's most advanced countries in the R&D and application of marine renewable energy by 2020.

CONCLUSION

With more than 14,000 km of island coastline and 4.7 million sq km of sea areas, China is rich in renewable energy but sees very poor development. Developing marine renewable energy is a major step the Chinese government shall take to meet the climate change, realize energy conservation and emission reduction, foster strategic emerging industries and accelerate the transformation of economic growth modes. Over the past years, China has seen sound development of marine renewable energy, which is attributed to legislation, planning, policies and funding. In 2010, the Ministry of Finance set up a special fund for marine renewable energy to fund demonstration projects related to power generation based on marine renewable energy as well as experiments of key technologies. This greatly pushes forward researches on marine technologies at home and speeds up utilization of marine energy. However, China still lags far behind developed countries in power generation with marine renewable energy. To narrow the gap, the Chinese government must overcome the obstacles in technologies, funds, policies and institutions related to the development and utilization of marine renewable energy.

REFERENCES

[1] Wang C; Lu W. (2009). *Approaches to analyze marine renewable energy resources and reserve evaluation.* Beijing: China Ocean Press.

[2] Zhou S; Qin S. (2009). *Strategic research on ocean development.* Hangzhou: Zhejiang University Press.

[3] Li C; Liao W; Wang Y. Progress in the technologies for power generation with marine wave energy. *the Magazine on Equipment Machinery*, 2010 2,68-73.

[4] Jiang Q; Bao X; Han X. Overview of the researches on and development of marine renewable energy in China. *Ocean Development and Management*, 2008 12,22-29.

[5] Wang Z; Ren D; Gao H. (2011). *2010 Report on the development of China's renewable energy industry.* Beijing: Chemical Industry Press.

[6] China Institute for Marine Affairs. (2010). *China's ocean development report.* Beijing: China Ocean Press.

[7] Wang Z; Ren D; Gao H; et al. (2013). *Report on the development of China's renewable energy industry.* Beijing: Chemical Industry Press.

[8] Jiao Y; Liu Y. Current development and prospects of wave power generation. *China High Technology Enterprises*, 2010 12, 89-90.

[9] Duan F; Zhong S. Current development and prospects of power generation with ocean temperature difference energy. *Journal of Northeast Dianli University,* 2010 2, 72-77.

[10] Shi J. Development of China's offshore wind energy and research on tariff policies. *Energy of China*, 2013 6, 25-32.

[11] Gao Y; Chai Y; Li H; et al. Current development of technologies related to marine renewable energy and suggestions on countermeasures. *Renewable Energy Resources,* 2011 2,152-156.

[12] Xia D. (2012). *Annual report of marine renewable energy (MER) in China.* Beijing: China Ocean Press.

[13] Xiong Y; Wang H; Cui L; et al. Research on the development and utilization of China's marine renewable energy. *Ocean Technology*, 2009 3, 106-110.

[14] Editor. Five obstacles facing China's marine renewable energy industry [online]. 2013[cited 2013 4 11]. Available from: http://news.bjx.com.cn/html/20130411/428138.shtml

[15] Zhang L. Preliminary exploration into the development and utilization of marine energy in China. *Shipbuilding of China*, 2012 2,555-562.

[16] Lovdal N; Neumann F. Internationalization as a strategy to overcome industry barriers: an assessment of the marine energy industry. *Energy Policy*, 2011 39, 1093-1100.

[17] Yan Q. Development of the world's marine renewable energy and the inspiration for China. *Ecological Economy*, 2012 8, 33-38.

[18] Yan Q; Liu L. Marine renewable energy—important support for the sustainable development of China's coastal economy. *Forum of World Economics & Politics*, 2012 3,159-162.

In: Renewable Energy in China
Editors: J. Yuan, G. Luo and J. Chen

ISBN: 978-1-63117-610-4

Chapter 6

SMART GRIDS IN CHINA

Jiahai Yuan[1,*], Jiakun Shen[1] and Huiming Xu[2]

[1]School of Economics and Management, North China
Electric Power University, Daxing, Beijing, China
[2]Information and Telecommunication Branch,
State Grid Corporation of China, People's Republic of China

ABSTRACT

Smart grid is the future direction of power system development and it has aroused wide attention. It is also the physical infrastructure with which to integrate renewable energy into the future power system. In China, grid companies are the pioneers in developing smart grids. Propelled by strong demand, China has made encouraging progress in smart grid development, especially in aspect of ultra-high voltage transmission system. However, in other aspects like distributed generation, microgrid and intelligent demand management, the progress is little. In this chapter we analyze the policy, pilot projects, achievements and barriers of developing smart grids in China. We find that the lack of a clear national strategy is one of the main institutional barriers for developing a smart grid in China. The current industrial structure of power sector, or the vertical integration of power transmission with distribution and supply, is also another institutional barrier of developing distributed generation, microgrid and intelligent demand response in China. Finally we provide an outlook on smart grid development in China.

Keywords: Smart Grid; Policy and institutional aspect; China

1. INTRODUCTION

The power system holds a pivotal role in renewable-based low-carbon energy system [1, 2]. Smart grid is the basic platform which enables the renewable energy system.

* Corresponding author: Jiahai Yuan. E-mail: yuanjh126@126.com.

Smart grid (SG) can contribute to renewable-based low carbon energy system in three ways. First, SG can enhance energy efficiency by improving the operation of traditional power plants and power grids. Second, SG can facilitate the integration of renewable power generation in large-scale concentrated and/or small-scale distributed ways.

Third, SG can boost various novel innovations and applications in demand response and promote energy efficiency in customer side [3, 4].

Since SG was proposed by EPRI in 2002, worldwide researches and construction processes of SG have taken off and many countries have established their own comprehensive strategies, goals and pathways to develop it [5].

In the US, SG plan devotes to the development of safe, reliable and modern power grids, enhancement of demand-side efficiency and reduction of power supply cost under the background of aging power infrastructure. In Europe, the Super Smart Grid scheme aims at complementarily reconciling two approaches of renewable energy development, namely large-scale centralized way and small-scale, local and decentralized counterpart, to realize a transition towards a fully decarbonized electricity system. In Australia, renewable energy and energy efficiency are targeted in SG development and the focus is on intelligent meters and intelligent demand-side management. In Japan, the emphasis of SG is to build renewable-friendly power grids adapted to large-scale solar power deployment, so as to resolve the conflicts between small territory, energy resource shortage and economic development. In Korea, the emphasis of SG research is on the integration of power system with smart green cities.

SG is also being regarded seriously in China. Grid companies took the initiative in developing SG. In May of 2009, State Grid Corporation of China (SGCC) released its vision and developmental roadmap for building a Strong Smart Grid (SSG) [6]. China Southern Grid Power Corporation (CSG) proposed its vision to build a smart, high efficient and reliable green power grid in July of 2010 [7]. In November, 2009, Ministry of Science and Technology (MOST) released a special report addressing SG related technology research and development (R and D) in China [8]. Then in March of 2011in the 12th Five-year Plan (FYP) on National Economic and Social Development advancing Smart grids, enhancing the resource allocation capacity of power system and improving power supply reliability, were taken as the key tasks for delivering power system transition [9], which is a clear signal that Smart Grid has been included in China's national energy policy. Shortly after that, in March of 2012, MOST issued a specialized plan for the industrialization of SG technologies and proposed to build the Chinese version of SG in 2020 [10].

In parallel to the policy advancement, there are encouraging technical innovations and many pilot projects implemented by the two grid companies in China [11].

Developing SG is not simply an engineering or technical issue. The transition to smart and low carbon power systems represents multiple challenges, including institutional, economic and managerial ones. In this chapter, we will first discuss the Chinese perspective of smart grid and compare the difference with those of other countries. Next we will compare the differences between smart grid and the traditional one and identity the key areas of SG development. In turn we review the policy progress relevant to SG in China. Then strategic planning on SG in China will be analyzed and progress on China's SG development will be reviewed. Next we will identify the barriers to SG development in China, in a perspective of institutional analysis. Finally we will discuss on the prospective of SG in China and conclude the chapter.

2. Smart Grids and Stakeholders in China

2.1. Perspectives on Smart Grid in China

SGCC is the pioneer for developing SG in China. As early as May 21 of 2009, SGCC put forwards its strategic plan for building a Strong Smart Grid (SSG). Meanwhile SGCC established an ambitious target, namely establishing digital, automated and interactive power grids which are supported by ultra-high voltage (UHV) transmission network.

According to Electric Power Research Institute (EPRI) of SGCC, SG can be defined as "a new type of highly integrated power grid, which is the combination of modern advanced sensing and measurement technology, information technology, communication technology, control technology and physical power system". SG is strong, self-healing, compatible, economical and integrative [12].

Some experts adopt the conception of "interactive power system". A version by Chinese Academy of Science, argues that "interactive power system" is not simply the renovation of traditional power system with modern information technology [13]. According to it, the prevailing concept is unable to define the basic pattern of next generation power grid and is also unable to generalize the core of modern power system in China. Hence Wu states that "interactive grid" is not merely UHV, and it should become the primary goal of power system development in China.

Some experts define SG in other way. In [14] Yu points out that the prominent feature of SG is the bi-directional flow of information and power in so as to establish a highly automated and widely distributed network for energy exchange.

Also, for the purpose of real-time information exchange and nearly transient balance of demand and supply, it is necessary to introduce distributed computation and modern communication system into the power system.

2.2. Definitions of SG around the World

In contrast to the development of UHV in China, some European countries initiated the construction of distributed grid (DG) to spread the risks. In countries like the United States and Russia, the pathway of UHV is almost abandoned. In the United States and Europe, the priority of SG development focuses on the following aspects: improving customer service, supporting the grid access of distributed renewable energy, localizing load and power supply, advanced automation and distributed intelligence, flexible grid operation, service-oriented architecture, more reliable and safe power supply, and finally energy efficiency. Specifically, the definition of Grid2030 by the US department of energy (DOE) goes as "a fully automated power delivery network ensuring a two-way flow of electricity and information between the power plants and appliances and all points in between" [15]. The definition by US EPRI (IntelliGrid) is "one electrical power system made up of many automatic transmission and distribution systems in coordinated, effective and reliable ways to complete all of the power grid operations in such a way: self-healing and adaptive; interactive with consumers and markets; optimized to make best use of resources and equipment; predictive rather than reactive to prevent emergencies; distributed across geographical and organizational boundaries; integrated, merging monitoring, control, protection, maintenance, EMS, DMS, marketing, and IT; and more secure from attack" [16]. The definition by European Technology Forum goes as "an electricity network that can intelligently integrate the actions of all users connected to it - generators, consumers and those that do both - in order to efficiently deliver sustainable, economic and secure electricity supplies" [17].

From the above definitions, it is clear that the European viewpoint is based on the European Union's desire to standardize the power grids in member nations and to interconnect them into a European Grid, while the definitions of US DOE and EPRI are also based on the examination of American power industry and point out the future direction of development.

From different definitions of Smart Grid by Chinese institutions, it is clear that SGCC prefers developing UHV on the basis of existing physical infrastructure. Informatization, digitization, automation and interaction, all these are the main features of future Smart Grid, but are described as just the characteristics of UHV power grid in the definition of SGCC. As a matter of fact, "next generation Grid technology", instead of Smart Grid is used in its definition. In a sense, SG is deliberately ignored. Perhaps the concept of "interactive grid" is more accurate and consistent with the global consensus on the future of power system.

2.3. Stakeholders of SG

For a grid company who desires to strengthen its vertical integration in the power system, SGCC naturally tends to take UHV as its priority. The concept of "interactive power grid" is proposed for the purpose of seizing advantage on the standard of SG for China in global arena [13], while [14] emphasizes the importance of real-time information in process of the smart grid operation from the perspective of practical work. As SG covers a wide variety of functions and applications, different stakeholders have different concerns on its development (table 6-1).

Table 6-1. Stakeholders and their concerns on SG

Stakeholders	Major concern
Central government	Solution for the economy-energy-environment puzzle; Deployment of renewable energy; revitalization of equipment manufacturing industry and creation of strategic new industry
Local government	Adequacy for local power supply; Promoting the development of local economy
Equipment manufacturers	Establishing the technology standard as soon as possible; seizing market share
Grid companies	Integrating renewable energy into the grids; Pricing mechanism and cost recovery; Real-time information of customer demand
power consumer	Stable power supply; Cost efficient of power supply
R and D institution	Holding advantageous position in setting up the world industry standards
Peripheral industries (ICT, **electric vehicle, etc.**)	Promoting the integration of multi-network; Promoting the innovation and deployment of new technologies in power grids; Supporting the development of EV industry

3. Comparison of Traditional and Smart Grid

Though smart grid is defined differently around the world, it is clear that there exists vast difference between it and the traditional power system. Generally speaking, SG is efficient, reliable, interactive and compatibly open. In table 6-2 their differences are compared in terms of efficiency, reliability, compatibility, interaction and openness.

Technology advance must be realized in different aspects of power generation, transmission, substation, power distribution and utilization etc., in order to make the theoretical difference into reality (table 6-3).

4. Policy Progress Relevant to Smart Grids in China

The Ministry of Science and Technology (MOST) formulated a special plan for key SG technology industrialization projects during the 12th Five-Year-Plan in 2012 [10]. According to the plan, SG will step into the overall construction in a few years.

Table 6-2. Comparison of smart grid and traditional grid

Feature	Traditional Grid	Smart Grid
Efficiency	The integration of operational data and asset management is bad.	Efficient operation: expand data acquisition and pay close attention to accident prevention to reduce negative impact on customers and improve operational efficiency; Asset optimization: introduce advanced information and monitoring technology to optimize the usage of resources and equipment, lower operation and maintenance costs.
Reliability	The ability of self-healing is poor and it can't cope with the natural disasters well.	Self-healing: continuous assessment can detect, analyze, respond to and even restore power equipment or abnormal operation in local networks; Security: effectively resist the damages of external attacks from both physical system or computer systems
Compatibility	It is designed mainly for concentrated power generation and is not adapted to distributed energy.	A large number of distributed renewable energy sources can be easily accessed to the grid.
Openness	Enforce strict admittance to equipment.	Allow free access to the grid and interaction between equipment and the grid.
Interactivity	It is unilateral and customers have to accept whatever the service.	Inform customers and encourage users to participate in power market. Customer is no longer a passive user and has power to choose among service providers.

Table 6-3. Key aspects of developing smart grids in China

Aspect	Requirements	Key Technologies
Generation	Raise safe operation of power system by a substantial magnitude; Promote large-scale deployment of renewable energy; Promote R and D on utility-scale energy storage technology to adapt to the rapid development of intermittent power supply.	√ Coordination technology between the grid and the regular plants √ Integration and control technology for renewable energy √ Large-capacity energy storage technology
Transmission	Improve the transmission capacity greatly and reduce transmission cost; Optimize the operating conditions of transmission network; Improve the stability level of power system and promote system interconnection; Realize condition assessment, fault diagnosis and online status maintenance and risk early warning.	√ GPS positioning, advanced intelligent detection and inspection technology √ Intelligent diagnostic analysis and decision technology √ Condition based maintenance, life-cycle equipment management and intelligent disaster prevention technology.
Transformation	Improve system stability and reliability, transmission capacity, as well as the health level; Provide information for smart dispatch of power system; Enhance the level of substation assets management and operation	√ Intelligent substation automation technology √ Integration of online testing and intelligent diagnosis √ Intelligent detection device and automatic verification device
Distribution	Enhance the reliability of power supply, system efficiency and terminal power quality; Realize distributed generation, energy storage and micro network interconnection; Coordinate operation optimization, efficient interaction of demand side management, and realize integration management of distribution assets	√ Power distribution automation system and integrated intelligent distribution network control technology √ Distribution dispatch and information system √ Distributed generation/storage and micro network access and coordinated control technology
Utilization	Enhance the quality of power supply service; Improve terminal energy efficiency and grid operation efficiency; Raise the proportion of clean energy in terminal energy consumption; Promote energy conservation and emissions reduction	√ Real-time data-mining and management of customer consumption patterns and behaviors √ Intelligent community/buildings √ Intelligent customer service system √ Customer-side distributed power and energy storage system

Aspect	Requirements	Key Technologies
Dispatching	Realize networked data transmission and visualization of operation monitoring; Realize dynamic safety assessment and refined dispatch; Realize automatic operation control and optimized generation-grid coordination	√ Energy management system √ Wide area measurement system (WAMS) √ Dynamic stability test and early warning system; √ Schedule management system
Information platform	Realize whole process information integration; Realize resource optimization and risk management on grid level; Improve interactions among grid operator and key stakeholders; Realize smart decision in big data system; Provide communication carriers for Internet of Things (IOT) and expand its applications	√ Modern information technology √ Big data platform for the whole process of power supply chain √ Integrated communication platform √ Business collaboration and interoperability platform

Ever since 2006, a series of policies, including *Renewable Energy Law, Pilot Management Scheme for Renewable energy pricing and cost sharing, Special Fund for Renewable energy development (Interim Measures)* have been issued and provide strong support for renewable energy development in China. Meanwhile, they also promote the development of SG in China. At the national level no specific strategy or policy has been designed for SG. SG is an extremely complex system, including generation, transmission, substation, distribution, consumption, power dispatch, and information platform. All policies in these areas are also related to SG. We therefore summarize the related policies in a chronicle way (table 6-4).

In addition, the trend that SG will integrate with the Internet of Things (IOT) is also becoming increasingly apparent, along with the development of IOT.

It can be easily reasoned that the government prefers to support the development of new energy and equipment manufacturing industry.

The main purpose is to promote the economy development. While for the power system, on the national level no clear direction has yet been pointed out.

All the related policies issued in recent years are concerned with solving the issues of deploying renewable energy and energy conservation in the current power system, but are not for the future development of SG. Lack of clear national strategy and integrative policy is the leading obstacle.

5. Strategic Planning on Smart Grids in China

Unlike in other countries where government plays a leading role in the development of SG, in China grid companies have more important role. There are only two grid companies in China: SGCC and CSG.

Table 6-4. Policies relevant to smart grid development in China

Time Issued	Policy	Main content	Relation with SG
January of 2006	Renewable Energy Pricing and Cost-sharing (pilot management scheme)	Stipulate two kinds of pricing mechanisms: government direct pricing and government guidance pricing. The guidance prices will be determined based on franchise bidding. The difference between RE price and regional de-sulfurized coal power will be shared among all customers.	Related with the price of renewable energy connected to power grids.
May of 2006	Special Fund for Renewable Energy Development in Buildings (interim measure)	Central government will allocate special fund to support renewable energy development in buildings.	Related with the deployment of new energy in SG.
November of 2006	Management Measures on Auxiliary service in Power System	Stipulate that power plants will provide two kinds of ancillary services: basic and paid ancillary services. It also indicates the provision and invocation, measurement and assessment, compensation and supervision related with auxiliary service.	Related with grid-connection of renewable energy in SG.
August of 2007	Energy-saving Power Generation Dispatching (Trial)	Stipulate energy saving as the principle of dispatch. Put renewable energy on the priority position in the schedule plan. Optimize dispatch in provincial and regional levels to minimize energy consumption and pollutant emissions.	Scheduling aspect of SG.
October of 2007	Admittance Management Rules on New energy vehicles	Provide preferential measures for the production of EV, including management way, production resources, technology and others.	Associated of EV /new energy storage applications in SG.
August of 2008	Special Fund for Wind power equipment industry (interim Measures)	Provide manufacturers with Subsidies for the first 50wind turbines (verified new model). The subsidy can only be used for R and D activity.	Related with new energy equipment manufacturing/ renewable energy deployment in SG.
January of 2009	Demonstration and Commercialization Financial Subsidy Fund for Energy-saving and new energy vehicle (trial)	Encourage usage of EV in public service in selected 13 cities. Provide subsidy for EV customers.	Associated with EV/new energy storage applications in SG.

Time Issued	Policy	Main content	Relation with SG
March of 2009	Finical Aid Fund for Solar Energy Applications in Buildings (trial)	Provide subsidy to solar power generation in buildings and the subsidy will be subject to adjustment according to technology progress.	Related with the deployment of distributed energy in SG.
May of 2009	Equipment Manufacturing Industry Restructuring and Revitalization Plan	Declare objective including GW-scale nuclear power unit, new energy power generation equipment and others to achieve independent manufacturing.	Related with new energy equipment manufacturing connected to the SG.
July of 2009	Golden Sun Demonstration Project	Arrange fund from renewable energy fund to support PV applications.	Related with new energy equipment manufacturing connected to the SG and distributed energy in SG.
July of 2009	Reform of Wind Power Pricing Mechanism	Set benchmarking price for wind power into four resource areas.	Related with renewable energy connect to the SG and pricing mechanism.
April of 2010	Renewable Energy Law (amendment)	Stipulate full protective purchase of renewable generation.	Related with renewable energy distributed generation and grid connection to the SG.
September of 2010	Fund Management Rules for Clean Development Mechanism in China	Establish national CDM fund as a response to climate change.	Related with new energy applications.
August of 2011	Reform on PV Pricing Mechanism	Stipulate direct government price for PV power generation.	Related with distributed energy in SG and pricing mechanism.
November of 2011	Instruction on Implementing Ladder Pricing Mechanism in Resident Customers	Reduce cross-subsidy and encourage energy conservation in resident power consumption.	Related with pricing mechanism in SG.
July of 2012	Pilot Project Management for Demand-side Management in Cities	Provide subsidy to cities for implementing DSM	Associated with customer response in SG.
December of 2012	Management Rules on the Special Fund for Strategic emerging industry (trial)	Establish special fund for promoting the development of strategic emerging industry.	Associated with the SG industry development funds.

These two companies take charge of almost all power transmission, distribution, dispatch, and customers service activities in China. The SGCC is a backbone state-owned enterprise covering 26 provinces, municipalities and autonomous regions, while the CSG serves other five provinces in the south, including Guangdong, Guangxi, Hainan, Yunnan, and Guizhou. As they serve different regions, they don't need to compete with each other.

5.1. Grid Companies

5.1.1. SGCC

As the largest grid operator in China, SGCC is the main promoter and developer of smart grid applications in China. SGCC launched a comprehensive planning for smart grid in 2010. According to the planning, Strong Smart Grid, which is backboned by UHV and coordinates the development of power grids at all level, will be the development direction of SGCC. A three-stage detailed plan is formulated to guide the development [18].

The emphasis of the first stage (2009-2010) is R and D on technical innovation and pilot projects. Now, the task of the first stage has been accomplished, with a plan, two fundamental construction programs, 21 demonstration projects, and 10 specific research projects finished. In the second stage (2011-2015), smart grids in China should be largely promoted.

Specifically, the construction of UHVDC and power distribution network would be accelerated. During the stage, the technical standard system for SG should be established and perfected. Also, major advance should be made on key technologies and equipment and they would be widely used in practice. In the last stage (2016–2020), the resource distribution ability, stability, interactivity between the power grid and the consumer should be enhanced, and a strong smart gird will be finally built. It is expected by the plan that at that time the integration of clean energy and its coordinative control should be significantly improved and utility-scale energy storage should be widely used. 100 GW wind power and 20 GW solar power should be connected to national power system and the system should be able to optimize more than 400 GW generation resource. The operation efficiency of the power system should also be improved significantly and the line loss would be lowered to 5.8% in 2020. Meanwhile, EV supporting and smart building infrastructure should be established and various novel value-added services would emerge. At that time various new business models around power grids would be established and widely accepted, thus creating a huge market.

5.1.2. CSG

Another power grid operator, CSG, has also put forward its overall objectives and principles concerning smart grid development [19]. CSG proposed to create a smart, efficient and reliable green grid, to promote resource-conservation and build an environment-friendly society.

A two-stage planning is proposed by CSG for the development of SG. The first stage (2010-2013) is planning, research and demonstration. To solve the problems in generation, transmission, substation, distribution, consumption, and dispatching, CSG would focus on R and D of key technologies and actively participate in the Key Technology R and D program initiated by MOST. The second stage (2012 and after) is the stage of demonstration and application.

5.2. NDRC

As the most powerful ministry in China, NDRC plays essential role in the development of any industries by formulating industrial planning, approving investment projects and providing fiscal support. In 2012 NDRC issued a special plan for "the industrialization of Smart Grid key science and technology" [20]. The plan stipulates the overall objective of developing SG in China as 1) acquisition of key technologies such as large-scale access of renewable energy, energy storage, and smart distribution, etc.; 2) formulation independent technology system and standard system for SG and establishment of an integrated supply chain; 3) and finally completion of the construction of modern smart grids in China. The plan also proposes to deploy demonstration projects and industrial training to realize the targets, including 20-30 projects for component technologies, 3-5 comprehensive projects, 5-10 Smart City projects, and 50 smart park projects.

It is worth noting the demonstration projects are provided by two grid companies and in this point, the NDRC plan intersects with the company-level plans.

5.3. Brief Comments

From the review on the relevant plans, we can have some interesting observations. On ministry level, the interest of NDRC is focused on the manufacturing industry and independent intellectual property rights. While for SGCC, UHV-based smart grid is the priority and SGCC hopes to take the leadership in China's SG development.

Therefore, SGCC formulates a very comprehensive and detailed plan for it. CSG, as a late comer, pays more attention on the development of relevant technologies and has a close eye the worldwide progress, but before the issue of explicit national policy will not arrange specific plan for it. It is also worth noting that even though manufacturing industries, renewable energy developers, EV industry and terminal customers are important stakeholders of Smart Grid, their demand can hardly be expressed in existing plans.

6. Progress of Smart Grids Development in China

6.1. Generation

In recent years, there is a rapid growth in China's power generation capacity and a significant improvement in the technology level of the power generation equipment. Clean energy generation develops rapidly and China becomes the largest wind power developer in the world. But thermal power, especially coal power still holds the leading position in the generation mix.

So there is still a long way for China to diversify its generation mix and cut reliance on fossil power. Also, due to the geographical conditions and natural resources restrictions, the installation of hydropower, pumped storage, gas and others power sources which can follow the demand quickly and provide auxiliary services to the power system, is insufficient and largely hinders the large-scale integration of renewable energy.

Table 6-5. Progress in generation sector

Key aspect	Project deployment and progress
Large scale integration of wind power	Formulate plan for developing 10 large-scale wind power bases (ranging between 3 and 10 GW) and the plan has come into effect.
Key technologies for the coordination of conventional power plant and power grid	Formulate a series of national technical conditions and test standards for the coordination between power grids and plants.
Grid connection, control and operation technologies for renewable energy power generation	In the end of 2011, the first pilot project on distributed generation, energy storage and micro-grid was accepted by SGCC and put into operation.
National wind power and solar power research and test center	On January of 2011, the National Research and Test Center for Wind Power was put into operation by ERI, SGCC. On July of 2012, the National Research and Test Center for Solar Power was put into operation by ERI, SGCC.
National demonstration project of wind/solar energy storage project	On December of 2012, first demonstration energy-storage project was put into operation by BYD corporation.
R and D and application of large-capacity energy storage equipment	Large-scale energy storage system is listed as one of the nine key major technologies in Smart Grid industrialization projects by MOST but no breakthrough has been made yet. Pumped storage hydropower station is taken as the priority presently.

Table 6-5 provides information on progress in generation sector, including coordination technology for conventional power plant and power grid, connection, control and operation technology for renewable power generation, energy storage, and capacity building for the large-scale integration of wind and solar power. It is evident that China has established national test center for wind and solar power and made smooth progress on the development of large-scale wind power base. Pilot project for distributed generation, energy storage and micro-grid has been implemented but the breakthrough on large-capacity energy storage is yet to be made.

6.2. Transmission

SGCC has comprehensively grasped the core technologies of UHV transmission system and developed the cutting-edge AC (1000kV) and DC (±800kV) UHV equipment as well as the test system, which effectively improve the safety and transmission capacity of the power grid. Table 6-6 provides information on the overall progress in transmission aspect. It is evident that China has made remarkable progress on FACTS and flexible HVDC technology R and D and application, and grid companies have already achieved the practical engineering on these technologies and integrated them into their operation practices.

Table 6-6. Progress in transmission sector

Key aspect	Project deployment and progress
FACTS application	By 2012, SVC, FSC, TCSC, CSR and STATCOM devices have been successfully used in practical projects. Presently the R and D direction focuses on fault current limiter (FCL) and other advanced equipment.
Flexible HVDC	In July of 2011, the first flexible HVDC transmission project in Asia was commissioned in Shanghai.
Intelligent inspection of power transmission lines.	Intelligent GPS patrol and helicopter patrol have been tested and proved the feasibility. The first set of unmanned helicopter patrol equipment has been made by SGCC in China. Various patrol robots have been developed by Chinese Academy of science (CAS), Wuhan university and other institutions.
Transmission line condition assessment and maintenance technology	The safety assessment on the components of transmission system and the corridor has been completed. Condition based maintenance is now widely implemented in China.

Table 6-7. Progress in transformation sector

Key aspect	Project deployment and progress
Construction and renovation of smart substation	SGCC has already completed the renovation of 1000 substations, and plans to build another 5100 smart substations and renovate 1000 substations during the 12th FYP period.
Intelligent substation operations	Reengineer the production process for smart substation and introduce the standardized mode of "dispatch and control center plus operation station" in most of the power supply companies.

However, the coordinated planning and development between renewable energy (especially wind power) and power grid is still problematic and partly results in serious curtailment of wind power in north China.

6.3. Transformation

Renovation of power system, condition based maintenance, and comprehensive automation for substation have been widely implemented in China. Now China is the leader in field of substation automation and has realized domestic manufacturing of the key equipment with proprietary intellectual property rights. All the newly constructed substations in China are equipped with integrated automation system and most of the existing substations have been upgraded. In all the substations of 110kV and above, function of "four-remote", viz., remote-metering, remote-signaling, remote-control and remote-regulation has been applied. Also, China has made substantial progress in reengineering the business process to adapt to the improvement in transformation technology (Table 6-7).

6.4. Distribution

Automation technology for power distribution has been deeply researched and preliminarily applied in the power system in China. Integration of distributed renewable generation on the distribution network is researched with remarkable results. In part of cities, various systems, including geographic information system (GIS), production management system (PMS), fault management system (FMS) and work management system (WMS), and implemented with the distribution monitoring system (DSCADA), customer management system (CMS), and enterprise resource planning (ERP) system have been developed and put into operation. In a word, the public supporting platform for handling distribution business has formed. However, the overall power supply capacity of distribution network and its reliability is still poor. Distribution automation has just covered about 9% of the distribution system and is often at idle state. Communication information network on distribution side is still in the early development stage. The applications of energy saving technology in distribution network are limited. The R and D and applications on distributed system and micro-grid is also insufficient. Table 6-8 presents an overview on the progress in distribution sector.

6.5. Utilization

Both SGCC and CSG have standardized their customer service system and business process in a practical way.

Research on information collection system for power customers is widely implemented and in part SGCC regions concentrated metering system is experimented. In CSG, metering automation system is been implemented with effort. Demand side management (DSM) has been widely deployed and receives considerable effect on energy conservation.

Table 6-8. Progress in distribution sector

Key aspect	Project deployment and progress
Distribution automation Integrated support system for distribution network control	In major cities the structure of distribution network is approaching mature. The primary equipment that can satisfy the requirement of distribution automation are technically mature. The terminal devices of distribution automation are technically mature. Various communication and host station systems are gradually applied in practical projects.
Construction of distribution information system	Intelligent decision support system for distribution network planning is in the process of comprehensive popularization.
Integration and control of distributed generation, energy storage and micro-grid	Pilot projects on distributed generation and energy storage have been implemented. Some customer-side roof PV projects have been put into operation. On technical level, currently the R and D activities mainly focus on the optimization, modeling and simulation of micro-grid and distributed generation.

Related standards, regulations and rules have been developed for information-based marketing support and customer service systems.

Research on the key technologies for charging electric vehicles has been initiated and some charging stations have been established in metropolis as Beijing and Shanghai for research and test purpose. However, the intelligent two-way interactive service platform has yet to be established. The technical support system for customer information collection and intelligent energy service is yet to be developed. Key technologies and applications for smart community, smart building, and customer-side distributed generation and energy storage are still in the void. Detection and management on smart metering devices is yet to be deployed.

More importantly, the relevant standard system for smart power utilization is yet to be formulated. Table 6-9 presents an overview on the progress in this aspect.

6.6. Dispatch

SGCC has occupied international advanced level on the technology and equipment of dispatch system and has technical advantages in perspectives of protective relaying, safety automation, WAMS, on-line stability analysis and early warning, secondary system safety protection. While CSG has just developed its automation system for dispatch in recent years and has limited experience on it. But in the past years, SCADA/EMS is widely applied in the CSG and contributed to the improvement on dispatch automation. But for both SGCC and CSG, the technology development on power system dispatch lags behind the development of UHV and large-scale renewable energy base.

Table 6-9. Progress in utilization sector

Key aspect	Project deployment and progress
Customer information collection system	The system is undergoing comprehensive construction in urban regions in both SGCC and CSG. In the future, it will be applied in rural regions.
Smart building/ community	Application of intelligent building has been initiated in Beijing, Shanghai and other cities. SGCC has deployed a pilot smart community project (Sino-Singapore Smart Eco-city) in Tianjin.
Charging infrastructure for EV	SGCC: In 2009, the first charging station was developed in Shanghai and in 2010 the first operational station was put into operation in Nanjing. CSG: In 2010, two charging stations with a total of 134 charging piles were put into operation in Shenzhen. By 2011 310 charging stations with a total of 16184 charging piles had been developed in China and covered 80% of provinces. Shanghai municipal government has issued subsidy policy to encourage private investment in charging stations and promises to underwrite 30% of the investment.
Research and test center for smart power utilization	The center has been set up by EPRI of SGCC in 2012.

For example, the technology of on-line power system safety analysis and control is weak. The prediction and regulation capability for the output of intermittent resources as large-scale wind power and solar power is also weak. In perspective of energy-saving dispatch, there is also much space to be improved. Table 6-10 presents an overview on the progress in this sector.

6.7. Comprehensive Projects

Besides the development of these special projects, some comprehensive demonstration projects have been deployed in China, for example, the Sino-Singapore Eco-City Smart Grid Demonstration Project and the Shanghai Expo Demonstration Project.

The 2010 Shanghai Expo project is the first comprehensive demonstration project in China. The purpose of the project is to provide a window to the public on what smart grid can do for them. Also, it provide an opportunity for SGCC (and Chinese Government as well) to show its (China's) technical capability on power engineering to the global society. It includes 9 demonstration projects and 4 presentation projects. Demonstration projects includes applications such as intelligent substation, distribution automation, fault repair management systems, power quality monitoring, information collection system, energy storage systems, new energy access, smart building/home, and EV charging and discharging. The presentation projects include smart dispatch, smart transmission, information platform and visualization.

While the Shanghai Expo project is mainly for presentation purposes, the Sino-Singapore Eco-City project is for practical application and serves as a prototype on the future smart grid services in cities. It covers almost all the aspects of applications: intelligent generation with energy storage and distribute source, standardized and safe intelligent transmission with digitization line, intelligent Substation with digital information, networked communication platform, standardized information sharing, and advanced interactive applications, Intelligent distribution based on distribution automation and featured with self-healing, flexible and adjustable, two-way interactive service platform including information collection system, intelligent two-way interactive services platform, intelligent power utilization based on smart community/buildings and charging facilities, high-speed communication and information network, and visualization platform integrated with GPS and analysis functions.

Table 6-10. Progress in dispatch sector

Key aspect	Project deployment and progress
Communication network project	It is setting up fiber-optic network, to further improve the performance of information network, safety and reliability, etc.
Access network for low-voltage power system	100% fiber-optic network coverage is under construction in newly-constructed urban communities. And tri-networks integration is under the way.
ERP project	SGCC and CSG have completed the construction of Enterprise Resource Planning system in 2012.
Integration of information applications	The projects include whole-process energy management system, life-cycle asset management system, customer relationship management (CRM) and enterprise risk management system and are under way.

7. Institutional Barriers and Policy Implications

Transition to a smart and low carbon power system represents multiple institutional challenges. For one thing, renewable energy in different forms is distributed and intermittent in nature. For another thing, without technically feasible energy storage options, power system must be balanced in real-time. Therefore, historically operators of the "dumb grid" were primarily concerned with stable and reliable operation of the system. The grid was operated using a combination of command-and-control actions and hierarchical decision-making. Perhaps most importantly, the preferences of customers were largely left out of the planning and operations process in traditional power grids.

Integration of medium-and-small-scale renewable generation will face huge challenge.

As discussed above, smart grid based on renewable generation and energy storage technologies radically differs with traditional one. In China, renewable generation as wind and solar power is integrated into the system by building large-scale bases and then constructing transmission system to deliver it. In this way, wind farms are taken as bulk generation of coal-based units. However the integration of medium-and-small scale renewables into the grid means that the renewables will be connected in the distribution system or directly built on customers' houses/buildings. So first of all, level playground for small and private investors is needed to attract millions of prospective individuals into the power sector. Then regarding power grid as public-accessible public goods and ensuring fair grid-access is another priority to overcome this barrier.

Pricing mechanism is the biggest challenge for smart management of customers and innovations in demand side technologies. Though currently smart meter and the related information and management technologies are the focus of R and D and pilot projects, price reform is ultimately the final answer. For one thing, retail tariff must reflect the cost of energy service in consideration of the time served and quality requirement. For another thing, currently there is only wholesale generation price and final consumption tariff based on customer classification and grid companies' revenue is the difference between them multiplying by power supply to different customers. Under this situation, Grid companies have no incentive to deliver demand side management.

Also, electricity price policy has historically assumed too much economic growth and social stability function in China, instead of economic function to signaling for the scarcity of resource. So to reflect the cost clearly, there must be price catalogue of wholesale generation, distributed generation and micro-generation and the separation of generation, transmission, distribution and retail prices, which in turn, calls for a new perspective on the regulation of transmission and distribution network. Understanding that there is deep-rooted origin of the current tariff policy, pricing reform can be regarded as a touchstone to see whether Chinese Government is determined to the Scientific Development Perspective and power sector deregulation.

The integration of medium-and-small scale renewable generation as well as pricing mechanism reform also calls for the gradual redefinition of the power grids and restructuring of the utility companies. It would pose the most difficult institutional puzzle for developing smart grid in China. Currently, SGCC, the biggest national grid company is the first mover of smart grid. But the focus of SGCC is on UHV and long-distant transmission system, which can serve as a physical infrastructure to maintain and strengthen its vertical monopoly power.

However, the worldwide trend requires splitting the grid company into transmission operator, distribution operator and retailers, among which transmission and distribution operator will provide public goods to generators, retailers and customers. Without clear-cut separation of public goods and private goods, it is unlikely that smart grid can function as expected. The puzzle is that, the grid company would have no incentive to develop SG if it was destined to be dismantled. Therefore, a subtle design to synchronize the SG roadmap with power sector deregulation is necessary.

Finally, there is another challenge to draft a clear national strategy and roadmap for developing SSG when trying to put all the elaborated components together and into operation. Smart grid is perhaps the most complicated manmade system and involves at least electric equipment manufacturing industry, power industry, information and communication industry and other novel industries which provide component solutions. Some components, as smart substation, FACTS, HVDC, Wide-area monitoring and control technology are mature and available. Others, like smart meters, DG, micro-grid, next generation communication and control system, Internet of Things, are still in R and D or early pilot stage.

Therefore, drafting a clear national strategy and roadmap based on comprehensive appraisals of numerous component technologies and their launch pathways poses another puzzle for developing SG in China.

An accompanied issue is how to integrate the international endeavors with domestic efforts.

8. Prospective of Smart Grids in China

8.1. Investment on SG in China, 2011-2020

China's rapid economic growth is accompanied by swift growth of electricity consumption. It is projected that in 2015 total electricity demand will reach 5815 TWh and the maximum load will reach 1300 GW; in 2020 total electricity demand will reach 7800 TWh and the maximum load will reach 1790 GW [3]. Accordingly, based on the planning of SGCC and CSG, We compile the estimated investment on power in China during 2011-2020 in table 6-11 [6, 18].

According to the planning rate for investment in smart grid over total power grid investment and the detailed investment share, we compile China's SG investment into 2020 in table 6-12. We find that about 25% of SG investment is on transmission and substation, 23% is on distribution, 30% is on customer side and another 18% is on dispatch information, and communication.

Table 6-11. Estimated investment on power grid in China, 2011-2020

	2006-10	2011-15	2016-20
SGCC	1215	1500	1400
CSG	234	288	270
Total	1449	1788	1670

Note: 1 US$=6.3 RMB in 2012.

Unit: Billion RMB.

Table 6-12. Estimated investment on smart grid in China, 2011-2020

	2011-15	2016-20
Total power grid investment	1788	1670
Share of SG in total power grid investment	11.7%	12.5%
Total SG investment	209.2	208.7
Generation	3.35	3.34
Transmission	10.87	12.94
Smart substation	43.72	40.70
Distribution	45.4	48.4
Utilization	69.24	64.28
Dispatch	7.32	7.93
Information and communication	29.30	31.10

Unit: Billion RMB.

8.2. Benefits of Developing SG in China

With the expected development, SG will result in massive benefits. From a utility's perspective, a Smart Grid can be viewed as a means to further five primary goals: (1) enhance customer service; (2) improve operational efficiency; (3) enhance demand response and load control; (4) transform customer energy use behavior; and (5) support more utility energy efficiency investment [21]. From the perspective of the government or the whole society, developing Smart Grid can bring forth more energy conservation and GHG reduction than would otherwise be attainable, and promote employment and economic growth. In this subsection, the benefits of developing SG in China will be quantified in six aspects, viz., avoiding investment in power sector, reducing the operation cost of power system, resulting environmental benefits, bringing forth customer value added, promoting related industrial growth, and finally promoting employment and stimulating economic development.

Start with an avoided investment in the power sector. It is estimated that in 2020 the deployment of SG would effectively avoid 63 GW of coal power installation, or reduction investment in coal power plant of 220 billion RMB, which in turn would avoid annual fuel cost of 19 billion RMB.

Assuming that the investment rate between power generation and grid is 1:1.5, the avoided investment in power grid would amount to 330 billion RMB. For system operation, SG could promote efficiency improvement of the traditional thermal power plants and decrease coal consumption per KWh electricity generation by 5.8 grams, which would amount to annual energy conservation of 32 million tonnes coal equivalent (Mtce). For line loss, it is estimated that SG could reduce overall line loss of the power system by one percent point, which would amount to annual energy conservation of 25 Mtce in 2020. Turn to clean development. It is estimated that the development of SG would increase renewable energy (wind and solar power) installation at least by 25-30 GW in 2020, which could substitute 45 Mtce of primary energy consumption annually. SG can also promote the development of electric vehicles (EV). According to the industry data, EV will emit 12 CO_2 kg per 100 km mileage, while a regular car will emit 19 kg per 100 kilometers mileage. So it is evident that for the same 100 km mileage, EV can reduce CO_2 emissions by 7 kg.

Suppose that in 2020 there would be a total of 30 million EVs in China, with an annual mileage of 10000 km, the penetration of EV supported by smart grid could reduce CO_2 emissions by 21 Mt [18, 19].

SG can bring forth vast benefits to customers in term of reduced electricity bill. According to the estimate by SGCC and CSG, SG could bring forth a total of 4850 TWh electricity conservation in China during 2011-2020, which could amount to saving of 2790 billion RMB for customers [18, 19].

SG will also promote the development of related industries. Besides renewable sectors like wind turbine and PV panel manufacturing, sectors like EV manufacturing, ICT industry, smart appliances and smart meters etc. will benefit from the development of smart grid. Though it is difficult to quantify the economic value, it is certainly that SG will promote the restructuring of Chinese economy and contribute to millions of employment in these hi-tech industries.

CONCLUSION

China has made encouraging progress in smart grid development, especially in the aspect of ultra-high voltage transmission system. However, in other aspects like distributed generation, micro-grid and intelligent demand management, the progress is limited. Our review reveals lack of a clear national strategy and roadmap, as well as the grid companies' monopoly position and their incentive structure under the existing regulation environment are the main underlying factors contributing to the imbalanced development of smart grid in China. China must prudently promote power sector reform along with the development of smart grid. In this way, SG will have a bright future and contribute greatly to renewable energy development and energy conservation in China.

REFERENCES

[1] Pacala, S., Socolow, R. Stabilization wedges: solving the climate problem for the next 50 years with current technologies. *Science*, 2004 305(13), 968-972.

[2] James, H., Williams, et al. The technology path to deep greenhouse gas emissions cuts by 2050: The pivotal pole of electricity. *Science*, 2012 335 (53), 53-58.

[3] Yuan, J., Hu, Z. Low carbon electricity development in China-An IRSP perspective based on Super Smart Grid. *Renewable and Sustainable Energy Reviews*, 2011 15(9), 2707-2713.

[4] Tan, W., He, G., Liu, F., et al. A preliminary investigation on smart grid's low-carbon index system. *Automation of Electric Power Systems*, 2010 34(17), 1-5.

[5] SGCC. The roadmaps for smart grids of developed countries. *Journal of SGCC* 2, 2012, 73-75.

[6] SGCC. *Smart Grids Development of SGCC during 12th Five-Year-Plan period*. 2009.

[7] CSG. *Study on the Strategy and Planning of Smart Grids Development of CSG*. 2010.

[8] MOST. *Report on Accelerating Technological Development of Smart Grid in China*. 2009.

[9] State Council of China. *The 12th Five-Year Plan for National Economic and Social Development.* 2011.

[10] MOST. *Special Planning for Key Scientific and Technological Industrialization of Smart Grid during 12th FYP Period.* 2012.

[11] Yu, Y., Yang, J., Chen, B. The smart grids in China-a review. *Energies*, 2012 5(5), 1321-1338.

[12] Hu, X. H. Smart grid-a development trend of future power grid. *Power system Technology*, 2009 33(14), 1-5.

[13] Wu, J. D. Innovative development of smart power grid and interactive smart grid in China. *Power System and Clean Energy*, 2009 25(4), 5-8.

[14] Yu, Y. X., Luang, W. J. Basic philosophy of smart grid. *Journal of Tianjin University*, 2011 44(5), 377-384.

[15] DOE. *"GRID 2030": A National Vision for Electricity's Second 100 Years.* 2003. Available from: http://energy.gov/oe/downloads/grid-2030-national-vision-electricity-s-second-100-years.

[16] IntelliGrid Program. Available from: http://www.smartgrid.epri.com/IntelliGrid.aspx.

[17] Smart Grids European Technology Platform. Available from: www.smartgrids.eu.

[18] SGCC. *The overall report on smart grids planning of SGCC.* 2010.

[19] CSG. *Research on the strategy and planning of developing smart grids in CSG.* 2010.

[20] NDRC. *Special plan for the industrialization of Smart Grid key science and technology.* 2012.

[21] EPRI. *The Green Grid: Energy Savings and Carbon Emissions Reductions Enabled by a Smart Grid, Palo Alto, CA.* 2008. Available from: http://www.smartgridnews.com/artman/uploads/1/SGNR_2009_EPRI_Green_Grid_June_2008.pdf.

In: Renewable Energy in China
Editors: J. Yuan, G. Luo and J. Chen
ISBN: 978-1-63117-610-4

Chapter 7

SUCCESSES AND FAILURES OF CHINA'S GOLDEN-SUN PROGRAM

Guoliang Luo *

School of Economics and Management,
North China Electric Power University, People's Republic of China

ABSTRACT

As export was impeded and products became unsalable, China's solar PV enterprises found it urgent to tap domestic market in 2009. As a result, the Ministry of Finance (MOF), the Ministry of Science and Technology (MOST) and the National Energy Administration (NEA) jointly released the policy about the Golden-Sun Program, deciding to provide investors with upfront subsidies prior to project construction with the special funds for renewable energy, with the aim to accelerate large-scale industrialization of PV power generation in China. During 2009-2012, more than 650 projects in five phases were approved as eligible for the Golden-Sun Program, involving a total capacity of 5,930 MW and 28 billion yuan of subsidies from the MOF. The Golden-Sun Program is a transitional policy implemented as an emergency plan and generated amid the severe economic and financial situations both at home and abroad. It plays an active and important role in quickly igniting China's PV market. But undeniably, the policy has institutional weaknesses, and the poor regulation leads to worse-than-expected quality and effects of engineering projects. Implementation of the Golden-Sun Program offers rich experience for China in its exploration into the development model, policy system, management system, feed-in power tariff and technical standards for distributed PV power generation. This paper suggests four indicators used to assess the development of China's PV industry, i.e., policy system design, efficiency, administrative institution and time frame/development period, and assesses the success and failure of the Golden-Sun Program using the said indicators.

Keywords*:* Golden-Sun Program, investment subsidy, Feed-in Tariffs (FIT), regulation, grid connection

---

* Email: lgl365365@126.com.

1. INTRODUCTION

China's solar PV power generation started in the 1970s and entered a period of steady expansion in the middle 1990s. In the early 1990s, PV power was mainly applied in communications and industrial fields and remote areas. In 1995, more than 40 PV power stations were built at the county, township and village levels, and about 150,000 sets of household PV power supply systems were set up [1].

In 2002, the Chinese government initiated the Township Electrification Program, in a bid to provide electricity for households in villages of seven western provinces and autonomous regions without power supply by using a mixture of photovoltaics and small wind power. The program greatly stimulated the solar PV market, with several solar cell encapsulation lines put in place across China that resulted in the rapid increase in the annual output of solar cells. In October 2003, the State Development and Planning Commission and MOST developed the Program on Development of Solar Energy Resources in the Next Five Years, and decided to implement the Brightness Program, raising 10 billion yuan to promote application of solar power generation technologies and achieving an installed capacity of 300 MW in nationwide solar power generation systems by 2005 [2]. In January 2006, the *Renewable Energy Law of the People's Republic of China* officially took effect, providing a legal basis for application of PV power generation systems at home. However, it is undeniable that China's PV application market is relatively backward with a rather small market capacity and the rapid development of the PV industry has been mainly driven by the international market.

Since 2000, Germany has taken the lead to introduce the feed-in tariff (FIT) policy mechanism, which greatly promoted expansion of its PV market. Other European countries followed the example of Germany and successively implemented the FIT mechanism, and therefore the PV market across Europe has seen a rapid growth. From 2000 to 2010, the world's PV market registered an average annual growth rate of 54.1%. Thanks to the driving force from the European market, China's PV industry has developed rapidly since 2004, with the growth rate exceeding 100% for five consecutive years. In 2007, China became the world's biggest producer of solar cells for the first time [3].

Pinched by the US financial crisis and European debt crisis, China's PV industry has seen a shrinking overseas market since 2009, making exports impeded and PV enterprises shut down. Given the economic and financial situations both at home and abroad, the Chinese government determined to increase policy incentives by implementing the Golden-Sun Program to accelerate the growth of the PV market at home. The structure of the chapter is clarified as: Section 2 is about methodology; Section 3 introduces the context for the launch of the Golden-Sun Program; Section 4 describes changes in the policies about the Golden-Sun Program; Section 5 analyzes the success of the Golden-Sun Program; Section 6 assesses the failure of the Golden-Sun Program, and Section 7 is conclusions.

2. METHODOLOGY

Morthorst (1999) identified five key areas as the most important indicators for evaluating the process of wind power development: these are investment certainty, effectiveness, market competition, and administrative demands [16].

- Investment certainty refers to the high level of certainty for independent investors who are guaranteed a fixed price for each kWh of power that is connected into the grid.
- Effectiveness is the degree to which tariff policy helps promote renewable electricity. It is important to determine the optimal level of subsidies for the most suitable investment in wind power [4].
- Market competition encourages wind power producers to evolve to adopt a market mechanism.
- Wind power development should have clear and transparent administrative demands that specify the extent to which the government should be involved in regulation or deregulation of tariff management.

After studying the policy experience about renewable energy power in Denmark, Germany and the UK, Judith (2007) found that the top factor affecting the development of renewable energy in Europe is the long-term policy guarantee for investors, and then the actual implementation of the policies [17].

Assessment of China's Golden-Sun Program must refer to the actual development of China's PV industry. Taking for reference the conceptual framework put forward by Morthorst and Judith, this paper proposes four indicators, i.e., policy system design, efficiency, administration institution, and time frame/development period. Reasons are as follows:

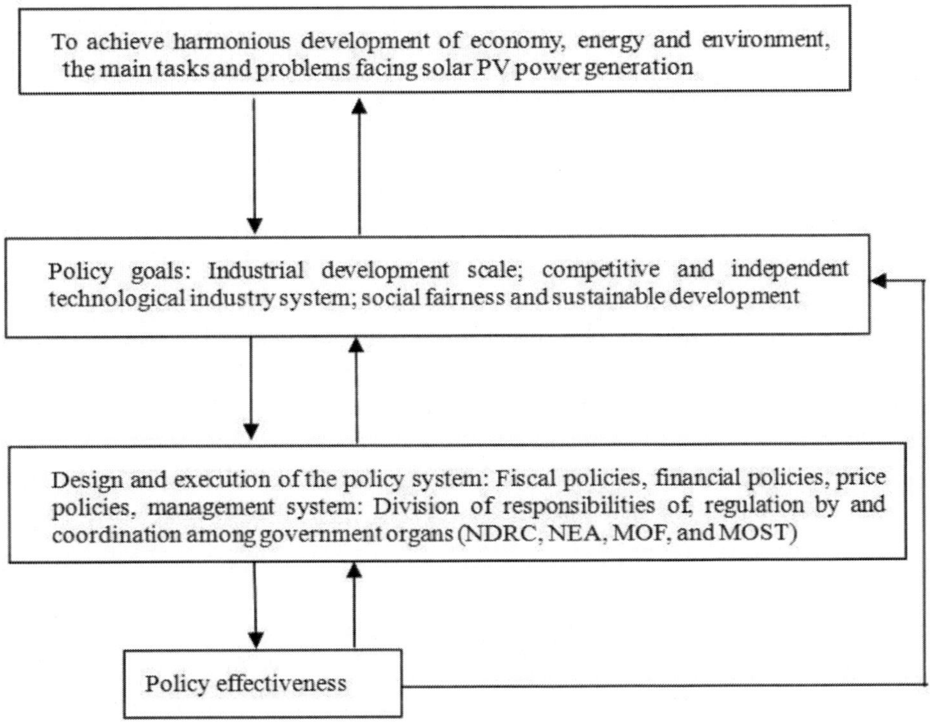

Figure 7.1. Formation of solar PV development policies and the effects.

Table 7-1. Roles of different incentives for the PV power generation industry at various stages

Incentives as policy instruments		Technical and commercial development stages of the PV power generation industry			
		Technology R&D	Project demonstration	Expansion stage	Maturity stage
Fiscal expenditure policies	Fiscal investment	O	●	●●	●●
	Fiscal interest discount	O	●	●●	●●
	Fiscal subsidies	O	●●	●●	●●
	Fiscal allocations	●●	●●	●●	●●
	Fiscal procurement	O	●	●●	●
Tax policies	Accelerated depreciation and investment tax credit	O	●	●●	●●
	Income tax reduction and exemption	O	●	●	●●
Financial policies	Policy loans	O	●●	●●	●
	Commercial loans	O	O	●	●

Note: O Almost no effects; ● Ordinary effects; ●● Notable effects.

In the middle and long term, China's main tasks in developing the renewable energy industry are to establish complete and coordinated incentive policies for the renewable energy industry and gradually expand and stabilize the renewable energy market [4, 5]. Whether the tasks can be successfully fulfilled depends on the coordination among renewable energy development on one hand, and economy development and environment protection on the other hand. The Chinese government defines the PV industry as a strategic emerging industry, which needs support from government policies at the initial stage because the investment costs are high and the market is at the growth stage. But whether the policy support is effective relies on the supporting system, the effective execution, and the coordinated and effective management system (See Figure 7.1). To comprehensively assess development of the PV industry, we also need to review the roles of different incentives as policy instruments at various development stages of the PV power generation industry (See Table 7.1) [6].

The *time frame/development* period is another essential factor that should be considered during policy evaluation. The suitability of incentive policy should be tested over a certain period of time or a development period. Long-term examination could reveal weaknesses and allow the introduction of appropriate strategic solutions in a local context. During the evaluation process, one should bear in mind whether the policy adjustment has been in effect long enough to judge the results domestically [7].

3. Context for the Launch of the Golden-Sun Program

3.1. Laws and Policies Regarding Renewable Energy

In January 2006, the Renewable Energy Law of the People's Republic of China officially took effect; Afterwards, China successively launched a policy package including the Long and Medium-term Plan for the Development of Renewable Energy and the 11th Five-year Plan for the Development of Renewable Energy [15] (see Table 7.2 for details), which not only specified the general goals for China's renewable energy development, but also set long and medium-range goals for some industries. To achieve these goals, the Chinese government also specified incentives for renewable energy power generation in the Renewable Energy Law, Trial Measures for Management of Allocation of Renewable Energy Power Tariff and Expenses and Interim Measure for Allocation of Additional Revenue from Renewable Energy Power Tariff. There are three main principles:

- Power grid enterprises must purchase the feed-in power generated from renewable energy in full amount;
- Power grid enterprises purchase the power generated from renewable energy based on the feed-in power tariff that combines reasonable costs and reasonable profits;
- The difference between renewable energy power tariff and conventional tariff is allocated in power grid across the country.

Table 7-2. China's laws and policies regarding renewable energy (2005-2011)

Laws and policies	Released on	Released by
Renewable Energy Law	February 2005	Standing Committee of the NPC
Trial Measures for Management of Allocation of Renewable Energy Power Tariff and Expenses	January 2006	NDRC
Management Regulations Related to Renewable Energy-based Power Generation	February 2006	NDRC
Interim Measures for Management of Special Funds for Development of Renewable Energy	May 2006	MOF
Interim Measure for Allocation of Additional Revenue from Renewable Energy Power Tariff	January 2007	NDRC
Long and Medium-term Plan for the Development of Renewable Energy	August 2007	NDRC
11th Five-year Plan for the Development of Renewable Energy	March 2008	NDRC
Notice on Issuing the Interim Measures for Management of Fiscal Subsidy Funds for Application of Solar PV Power in Buildings	March 2009	MOF
Notice on the Implementation of the Golden-Sun Program	July 2009	MOF, MOST, NEA
Notice of the NDRC on Improving the Policies Regarding Feed-in Tariff of Solar PV Power	July 2011	NDRC

In early 2009, China's policies regarding renewable energy power connection have basically taken shape. As for actual implementation, the tariffs and tariff mechanisms have been defined for the power generated from wind and biomass, but no tariffs or tariff mechanisms has been specified for PV power [8]. As the FIT policies have never been clarified, China's PV power generation market has developed slowly. Before 2008, PV power was mainly used in off-grid projects, such as rural electrification, communications and industries (See Table 7.3) [6].

Table 7-3. China's PV market share (%)

	2006	2007	2008	2009
Market segments	41.3	48.8	28.4	17.3
Rural electrification	33.8	20.7	22.0	11.6
Communications and industries	20.0	18.4	32.8	18.3
Scattered use of solar energy	4.8	11.9	16.6	18.8
Building Integrated PV	0.3	0.2	0.2	44.0
Total capacity (MW)	80.0	100.0	140.0	300.0
Total	100.0	100.0	100.0	100.0

3.2. Analysis of China's Incentive Policies for PV Power Generation

Since 2008, the Chinese government has introduced an incentive package for PV power generation to kick off domestic PV market. For eastern provinces with robust economic strength, the government offers additional subsidies for PV power generation projects with self-raised funds and implements local policies on fixed FIT to promote the development of local PV industry. Jiangsu Province is the most representative one of all.

In summary, there are two major types of incentive policies for PV power generation in China, i.e., subsidy policies for initial investment and policies regarding fixed FIT (See Table 7.4) [9]. The subsidy policies for initial investment are mainly applicable to small and medium-sized PV power generation systems, such as solar building integrated PV demonstration projects (Solar Rooftop Program for short); as for large PV power stations, the approval policy is adopted in which one FIT is for one project, or tariffs are defined differentially.

Table 7-4. China's main PV subsidy policies

Project types	PV subsidy policies
Solar Rooftop Program	Projects with an installed capacity of 50 kW or above; Subsidy standard: 20 yuan/W in 2009, and 17 yuan/W in 2010. Fund source: Special funds for renewable energy development
PV power FIT projects	Among the 100-plus feed-in PV power generation projects built nationwide in 2007 and 2008, the NDRC only approved four PV power station projects in two batches, including two in Shanghai, one in Inner Mongolia and one in Ningxia, with the FIT as 4 yuan/kWh. Fund source: Renewable energy power tariff surcharges levied nationwide (2 thousandth of a yuan/kWh)
Franchise PV bidding projects	In 2009, the NEA organized franchise bidding for the 10 MW PV power station in Dunhuang, Gansu Province, with the FIT as 1.0928 yuan/kWh and the franchise operation period as 25 years; In 2010, the NEA launched the second round of franchise bidding, resulting in 0.7288-0.9791 yuan/kWh of power tariffs in thirteen 20-30 MW projects in western China based on the principle of bid winning at the lowest price.
FIT project in Jiangsu Province	Target tariff of ground feed-in power stations: 2.15 yuan/kWh in 2009, and 1.7 yuan/kWh in 2010 Fund source: Special funds for PV power generation set up by imposing additional tariffs

The following problems can be discovered from the review of the main PV subsidy policies implemented in China:

1) Projects under the Solar Rooftop Program are relatively small

The Solar Rooftop Program is implemented on the basis of the Implementation Opinions on Accelerating Application of Building Integrated Solar PV released by the MOF and the Ministry of Housing and Urban-Rural Development. But the projects under the Solar Rooftop Program are relatively small in scale. In 2009, 111 projects with a total capacity of around 91 MW were approved as eligible for the Solar Rooftop Program, receiving 1.3 billion yuan of subsidies across China [10].

2) The franchise PV bidding prices are relatively low

Considering that franchise PV bidding is a major means for a country to learn about prices, the Chinese government intends to employ this means to promote the development of China's PV industry. The implementation of franchise bidding projects leads to a substantial decline in feed-in PV power tariff from 4 yuan/kWh in 2008 to 1.09 yuan/kWh in the first franchise bidding project in 2009. But among the franchise bidding projects, the principle of bid winning at the lowest price results in significant reduction of project returns by bidding enterprises (the internal rates of returns of some projects are even below 5%), which obviously goes against the sustainable development of the PV market.

3) The FIT policies are not in place

The delay in defining FIT leads to difficulties in identifying the period of investment returns by enterprises. With unclear returns, investors will find it hard to take interest in PV power generation projects, which objectively restricts development of feed-in PV power generation. Feed-in PV power generation must be fostered with a mixture of price means and market mechanisms, and China is in an urgent need to launch fixed feed-in tariff to push forward market development and offer market guarantee for the large-scale development of the PV industry.

3.3. Influence of the Financial Crisis on China's PV Industry

China's PV manufacturing industry has been expanded as pushed by the huge demand in European market. Pinched by the US financial crisis and European debt crisis, China's PV industry has seen a shrinking international market since 2009, with exports hampered and some PV enterprises shut down.

The financial crisis gives rise to signs of trade protection in some developed countries, which increases risks and uncertainties to China's PV power generation industry. The strong negative influence of the financial crisis on China's PV product manufacturing attracts great attention from the Central Government, which had set the strategic goal of making non-fossil energy representing 15% of the total by 2020. In 2009, the Central Government devised the strategic guidelines of "maintaining growth, adjusting structures and expanding domestic demand". Promoting the large-scale development of China's PV market is both required by energy conservation, emission reduction and improvement in China's energy structure, and an effective measure to reduce risks in domestic PV market as well.

Given the economic and financial situations both at home and abroad, the Chinese government determined to increase incentives for the PV industry, in the hope of seizing the strategic opportunities for the development of PV power generation, kicking off domestic PV power generation market and expanding the presence of China's PV power generation industry in the world.

The Golden-Sun Program was launched against the said backdrop.

4. CHANGES IN THE POLICIES ABOUT THE GOLDEN-SUN PROGRAM

4.1. Process for Implementing the Golden-Sun Program

On July 16, 2009, the MOF jointly with MOST and NEA released the Notice on the Implementation of the Golden-Sun Program, which pointed out that to promote the technological progress and large-scale development of PV power generation and foster strategic emerging industries, the Central Government shall allocate some special funds for renewable energy to support demonstration application of PV power generation technologies in various fields and industrialization of key technologies (the Golden-Sun Program in short) [15]. The Notice specified the main steps for implementing the Golden-Sun Program, as described in follows (Figure 7.2).

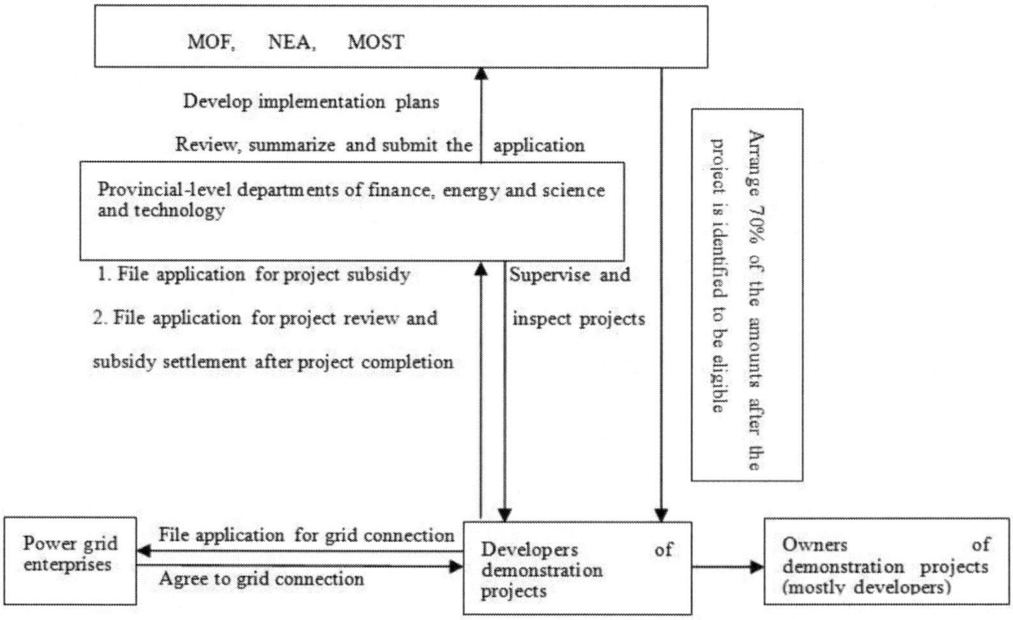

Figure 7.2. Process for implementing the Golden-Sun Program.

First, provincial-level departments of finance, science and technology and energy specify the areas, contents and schedules for the construction of demonstration projects in accordance with the implementation plans for the Golden-Sun Program, and jointly report them to the MOF, MOST and NEA for recordation.

Second, as for the projects included in the implementation plan, proposals and key equipment bidding shall first be completed. After local power grid enterprises give opinions that they agree connecting the power into grid, application for fiscal subsides can be filed. In the principle of territory, the application shall be submitted to the competent provincial-level departments of finance, science and technology and energy, who shall further submit the application to the MOF, MOST and NEA at the end of each February and August after organizing review of the application reports.

Finally, the MOF, MOST and NEA review the technical plans as well as materials about construction conditions and fund raising related to the projects. Afterwards, the MOF defines the subsidy amounts according to project investment and subsidy standards, arranges budgets based on 70% of the amounts, and settles the remaining subsidy funds according to actual project investment.

4.2. Areas Covered by the Golden-Sun Program and Implementation Measures

The Golden-Sun Program is an emergency policy launched by China to deal with the international financial crisis and promote the technological progress and large-scale development of domestic PV power generation industry. With changes in the PV market both at home and abroad, the policy measures have been adjusted slightly (see Table7.5) [15].

On July 16, 2009, the MOF, MOST and NEA jointly released the Notice on the Implementation of the Golden-Sun Program, deciding to accelerate the large-scale industrialization of domestic PV power generation by using fiscal subsidies, scientific and technological support and market incentives. The said three ministries and commissions planned to financially support PV power generation demonstration projects with the installed capacity not less than 500 MW in the next two to three years.

In 2009, 140 projects were approved as eligible for Phase I demonstration projects, to be completed in three years.

In 2010, large feed-in PV power stations are no longer covered by the Golden-Sun Program, and the rest of the supported areas were basically the same as those in 2009.

In 2011, the Golden-Sun Program highlighted support to the user-side PV power generation projects in economic and technological development zones and industrial parks. The subsidy standards were somewhat adjusted, i.e., key equipment bidding is no longer organized and equipment's compliance with standards is the key. Subsidies for initial investment are still adopted.

In the first half of 2012, the subsidy standards for Golden-Sun projects were adjusted.

In March 2013, the MOF decided to cease the review of new project applications related to the Golden-Sun Program; in May 2013, the MOF settled accounts related to the Golden-Sun projects.

Table 7-5. Main conditions, supported areas and measures for implementing the Golden Sun Program, 2009-2012

Document name, approval No. and date of release	Conditions and requirements for project implementation	Supported areas	Implementation measures
Notice on the Implementation of the Golden Sun Program, No. 397 [2000] of the Ministry of Finance, July 2009	The installed capacity of a single project is no less than 300kW; The developer has a total asset of no less than 100 million yuan, and the capital fund is no less than 30% of the total investment; Key equipment for demonstration projects, including PV modules, inverters and storage cells, are chosen through bidding.	1.Demonstration projects of PV power generation for user-side grid connection by large industrial and mineral enterprises, commercial enterprises and public welfare institutions; 2.PV and wind and solar projects in remote areas and areas without access to power; 3.Large-scale grid connection PV power generation projects in the areas rich in solar energy resources; 4.Demonstration projects for industrialization of key technologies in PV power generation.	PV power generation projects for grid connection are subsidized on the basis of 50% of the total investment, and the independent PV power generation systems in remote areas or areas without access to power are subsidized on the basis of 70% of the total investment.
Notice on Strengthening the Management of Construction of Golden Sun Demonstration Projects and Demonstration Projects of Application of Building Integrated Solar PV, No. 662 [2010] of the Ministry of Finance, September 2010	1.The installed capacity of the user-side PV power generation projects under concentrated and continuous construction is no less than 20 MW; 2.Key equipment is identified through unified bidding.	1.Support priorities shall be given to user-side PV power generation projects constructed by large industrial and mineral enterprises, commercial enterprises and public welfare institutions by using the established buildings and other conditions; 2.Independent PV power generation projects in remote areas and areas without access to power; 3.User-side PV power generation projects under concentrated and continuous construction in economic and high-tech development zones.	Supplement to provisions in 2009: Other expenses for the construction of demonstration projects are subsidized on the basis of fixed amounts with a standard of 4 yuan/W.
Notice on Effectively Conducting the Golden Sun Demonstration Work in 2011, No. 380 [2011] of the Ministry of Finance, June 2011	1.The installed capacity of the user-side PV power generation projects under concentrated and continuous construction is no less than 20 MW; 2.The installed capacity of user-side PV power generation projects is no less than 300 kW; 3.No unified biddings are organized for key equipment, and equipment complies with standards.	1.User-side PV power generation projects under concentrated and continuous construction in economic and high-tech development zones; 2.Demonstration projects of PV power generation for grid connection at the user side of large industrial and commercial enterprises and public welfare institutions; 3.Independent PV power generation projects in remote areas and areas without access to power;	Previous investment subsidy policies are cancelled, and the subsidy standard for demonstration projects is 8-9 yuan/W.
Notice on Effectively Conducting the Golden Sun Demonstration Work in 2012, No. 21 [2012] of the Ministry of Finance, January 2012	1.The total installed capacity of demonstration projects under concentrated application is no less than 10 MW; the installed capacity of user-side power generation projects under scattered construction is no less than 2 MW; 2.Key equipment is purchased independently.	Supplement to provisions in 2009: Other expenses for the construction of demonstration projects are subsidized on the basis of fixed amounts with a standard of 4 yuan/W.	The standard project subsidy is 7 yuan/W.

4.3. Effectiveness of Implementation of the Golden-Sun Program

From 2009 to 2012, a total of 655 projects in five phases were approved under the Golden-Sun Program, involving a total capacity of 5,930 MW (see Table 7.6) [10] and fiscal subsidies of 28 billion yuan on a cumulative basis [11] (See Figure 7.3).

Table 7-6. Approved projects under the Golden-Sun Program

Phase	Project number	Project capacity (MW)
Phase I in 2009	140	304.0
Phase II in 2010	46	271.7
Phase III in 2011	129	792.1
Phase IV in June 2012	155	1732.2
Phase V in November 2012	185	2830.0
Total	655	5930.0

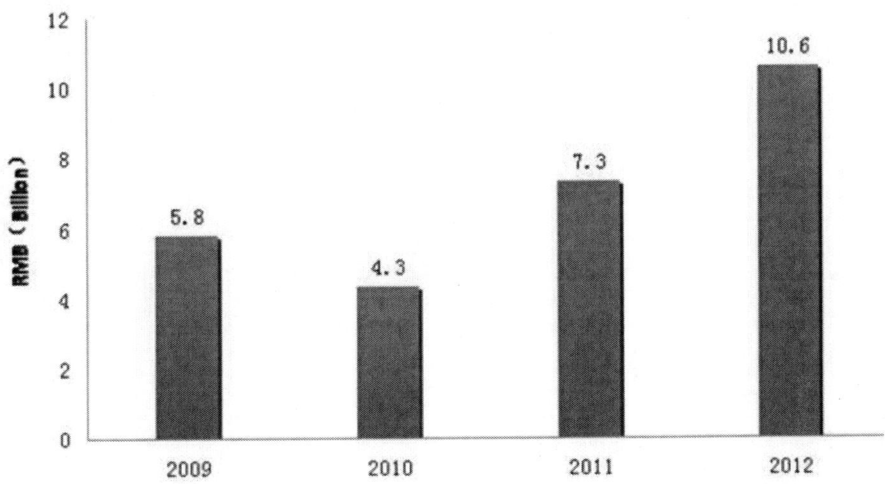

Figure 7.3. Subsidies to the Golden-Sun Program (2009-2012).

The Golden-Sun Program has pushed forward the rapid development of China's PV market as a whole, but newly increased PV capacities vary greatly with provinces (municipalities and autonomous regions)[19](see Figure 7.4)[12].

5. Successes of the Golden-Sun Program

5.1. At the Demonstration Growth Stage of the PV Power Generation Industry, Fiscal Subsidies Play a Notable Role in Promoting Growth of China's PV Market

Subsidies under the Golden-Sun Program refer to giving upfront funds that represent 50% of the total project investment to project investors in cash before project construction. In general, 70% of the subsidy funds will be paid prior to project construction, and the remaining 30% will be paid after project completion.

Before 2009, most commercial banks were reluctant to grant loans to PV power generation projects with unclear prospects, and private investors also did not prefer PV power generation projects with low rates of returns. Therefore, the PV industry had long been waiting for government policies about subsidizing PV users, which can truly enable effective connection between PV industrial investment and market consumption. But this obviously is a long process. The launch of the Golden-Sun Program greatly shortens such a process. Industrial investors and even financial investors immediately take interest in investment in PV power stations.

Before 2009, China's PV market developed slowly, with more than 95% of PV modules exported abroad. After 2009, domestic PV market grew by more than 100% annually, with the export share declining from more than 95% in the years before to 82.5% in 212, as shown in Table 7.7. [10].

Table 7-7. Shares of PV module export in China, 2006-2012

Year	2006	2007	2008	2009	2010	2011	2012
Export share (%)	97.5	98.2	98.5	96.0	95.4	87.9	82.5

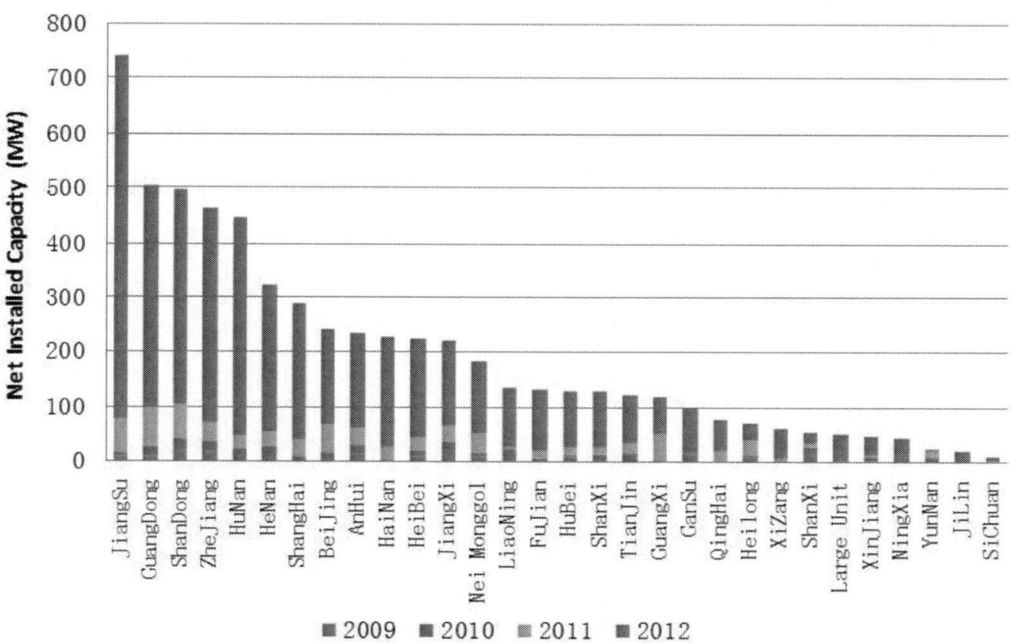

Figure 7.4. Increased capacities under the Golden-Sun Program in provinces (municipalities and autonomous regions).

As of the end of 2012, the projects covered by the Golden-Sun Program involved a total installed capacity of 5,930 MW (see Table 7-8) [12], wherein over 95% of such projects were distributed PV power generation projects. Therefore, China's distributed PV power generation made great headway, and the share of distributed PV systems in China's PV system market increased from 32.0% in 2011 to 36.4% in 2012. In particular, the Golden-Sun Program contributed to 80% of the development [10].

In a word, the Golden-Sun Program quickly kicks off and accelerates the development of China's PV market.

Table 7-8. Capacity of projects under the Golden-Sun Program, 2009-2012

Year	2009	2010	2011	2012	Total
Project capacity (MW)	216	275	692	4,747	5,930

5.2. The Golden-Sun Program Significantly Improves the Efficiency of Constructing PV Power Generation Facilities at Home

Due to implementation of the Golden-Sun Program and positive signals in the international market, China's PV market size and production scale have been expanded, the PV industry has been improved step by step, output of polysilicon materials has increased substantially and the cost declined, and prices of PV modules and PV systems have declined rapidly. According to a survey of PV module producers at the end of 2012, China's PV module manufacturing cost was about US$ 0.62-0.65. Compared with 2007, prices of PV modules declined by 87.5%, and system prices decreased by 83.3% (see Figure 7.5) [10].

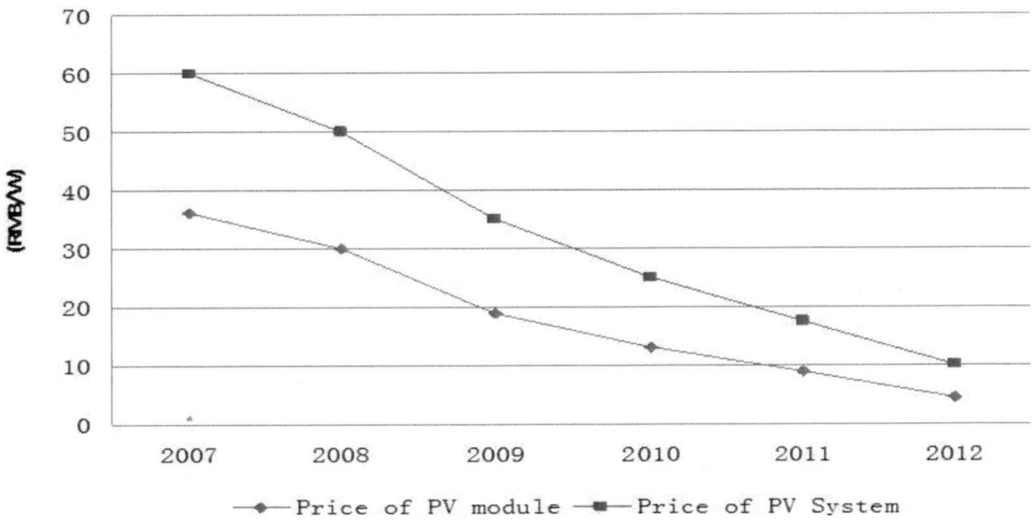

Figure 7.5. Changes in the prices of China's PV modules and PV systems.

5.3. The Golden-Sun Program Provides Clean Energy to Residents in Rural Areas with No or Insufficient Access to Power

The sparsely-populated and under-developed areas in western China are main beneficiaries of the Golden-Sun Program, which provides electricity to residents in remote mountainous areas in western China and brings clean energy to areas with no or insufficient access to power. Data shows that after three years of implementation of the Golden-Sun Program, local governments in Qinghai Province have applied for 46 projects, 15 of which were approved by the State and involved a total installed capacity of 33.4 MW and a subsidy of 552 million yuan from the Central Government. Benefiting from these projects, grassland herders gain better access to power in both living and production and share the fruits of technological civilization [18].

5.4. Management and Design of Distributed PV Power Generation Are Improved Significantly

Implementation of the Golden-Sun Program prompts a series of relevant standards and measures (see Table 7.9) [10].

Table 7-9. Standards and codes related to distributed PV power generation

Standards and codes	No.
Technical Code for Solar PV System of Civil Buildings	JGJ 203-2010
Code for Operation and Maintenance of Building Mounted Photovoltaic System	JGJ/T 264-2012
Code for Acceptance of Building Integrated PV Power Generation Systems	20111728-T-424
Technical Requirements for Connecting Photovoltaic Power System to Distribution Network	GB/T 29319-2012
Basic Acceptance Requirements for Grid-connected PV Systems	CGC/GF003.1-2009

To support implementation of the Golden-Sun Program, the State Grid changed the traditional management model in the power industry and released the *Opinions on Effectively Conducting Grid Connection of Distributed PV Power* in March 2013, which brings hope to China in breaking through the institutional obstacles to grid connection of distributed PV power.

Thanks to implementation of the Golden-Sun Program, China's design of distributed PV systems has improved significantly in terms of resource assessment and power generation forecast, building integrated PV design and solutions about orientations, shading and heat radiation, optimization and coordination of the tile angle of PV array, generating capacity, land occupied and wind resistance, design of PV systems and buildings, construction of building PV, design of grid connection systems, monitoring of distributed PV power generation as well as operation and maintenance of building PV systems. Meanwhile, China has fostered a number of PV system design and engineering construction talents as well as a number of highly professional and experienced enterprises.

6. FAILURES OF THE GOLDEN-SUN PROGRAM

6.1. Lack of Systematic Design and Effective Coordination in Policies

Stable and clear government commitment is the most key factor to policy success. The commitment includes price levels, effective terms, scope of participants as well as other supporting policies and implementation. The periods of protecting and supporting the PV power generation industry must be long enough to ensure its strong competitiveness compared with that of traditional energy. However, changeable policy orientations and poor implementation of supporting policies hurt investors. For example, Phase I Golden-Sun projects initiated in 2009 were planned to be completed in two to three years, but only fewer than half of the projects have been commenced. This is partly attributed to developers' intentional delay to reduce cost, but most importantly is the failure to implement supporting policies. Specifically speaking, the power generated cannot be connected to grid; user-side grid connection and use of self-generated power as encouraged by policies are hard to implement in practice and fiscal subsidies involve a long cycle to be put in place. The Golden-Sun Program is a policy means launched as an emergency response to the unbalanced PV supply and demand and outstanding problems in the international market, but it lacks coordination and continuity in implementation. Therefore, policy changes become usual [13].

6.2. Institutional Deficiencies in Policy Design

The Golden-Sun Program has two major institutional deficiencies: unclear and biased policy guidelines and unscientific subsidies. Details are as follows:

- The policy targets are not clearly confined to the users who generate PV power for self-use. In 2009, among the first batch of projects supported by the Golden-Sun Program, user-side ones accounted for 47%, large ground power stations 46%, and PV projects in the areas without access to power 7%.
- As for the installed capacity of a single power station, the lower limit rather than the upper limit is defined, and the lower limit has been raised year by year (see Table 7.5).
- The subsidy standards do not take into account the rapid decline in cost resulted from market competition, and are close to the cost of PV modules (see Table 7-10) [10]. The excessively high subsidy standards make owners of the Golden-Sun projects tend to purchase low-price equipment to reduce cost, which may do harm to the engineering quality. In 2010, the MOF announced the cancellation of 39 projects intending to make up the number, which involved a total installed capacity of 54 MW [11].
- Technological innovation and R&D are ignored. Although technological R&D is covered by the Golden-Sun Program and is offered certain incentives, it plays a very minor role when compared to the investment in PV power generation demonstration projects. As a result, China's PV industry has made certain headway in technologies, but the core technologies still lag far behind those of developed countries.

Table 7-10. Comparison between PV module costs and PV subsidies (yuan/W)

	2009	2010	2011	2012
PV module costs	19	13	9	4.5
PV subsidies	20	12	8- 9	5.5

6.3. Obstacles to Connecting User-Side PV Systems to the Grid

Although the State Grid released the *Opinions on Effectively Conducting Grid Connection of Distributed PV Power* in March 2013, which provided policy foundation for distributed PV user-side grid connection, the Golden-Sun Program was hampered by the PV system user-side grid connection during the whole implementation process (2009-2012), and the rate of grid connection of the Golden-Sun projects already put into operation was only 40%. This greatly reduces the effects of the Golden-Sun Program [14].

The common problem existing during implementation of the Golden-Sun Program is that local power supply enterprises are reluctant to let PV systems connect to the grid at the user side, but hope to manage PV power generation projects through public power stations. This leads to construction of booster stations under the Golden-Sun Program for concentrated grid connection from the transmission networks. But this conflicts with the user-side grid connection and use of self-generated power required by the Golden-Sun Program, i.e., the principle of operating the PV systems by offsetting quantities of power. That way, the values of PV power decline from the user-side power tariff (the average tariff of power for industrial and commercial use in eastern China is about 0.925 yuan/kWh) to the feed-in benchmark tariff of power generated by local desulfurized coal-fired units (only about 0.35 yuan/kWh), resulting in the total loss of project profitability. Moreover, compared with the power stations that use the power they generate, public power stations involve much higher construction cost, more complex approval procedures and longer period of time. So, the Golden-Sun Program has encountered the difficulty in user-side grid connection at the very beginning.

The obstacles to grid connection under the Golden-Sun Program are mainly attributed to the institutions. First, the Golden-Sun Program policies are mainly driven by the MOF, but the MOF fails to coordinate with power grid enterprises when devising the policies. This results in numerous difficulties in grid connection at later stages; and second, as the Golden-Sun Program encourages user-side grid connection and use of self-generated power and adopts the way of offsetting power quantities, owners purchase significantly fewer quantities of power from power grid enterprises, thereby affecting their economic benefits. Due to the lack of reasonable profits and incentives, power grid enterprises are somewhat against grid connection of the Golden-Sun projects.

6.4. Lack of Effective Regulation Leads to Poor Quality and Efficiency of Project Implementation

As the regulator of the power market, the State Electricity Regulatory Commission is responsible for regulation of power enterprises, power scheduling trading and fair opening of power grid by enterprises. However, regulation over PV power generation projects lacks supporting laws and regulations. The Notice on the Implementation of the Golden-Sun Program pointed out that local departments of finance, science and technology and energy are responsible for supervising and inspecting implementation of the Golden-Sun projects. But there are difficulties in executing the provision, as there are no detailed definition about who will dominate and coordinate the supervision and inspection and who will take the main responsibilities, resulting in the absence of regulation after the Golden-Sun projects are approved and implemented.

Even if local departments of finance, science and technology and energy are willing to supervise Golden-Sun projects, the huge supervision costs would deter them from coming forward. Projects need to be applied, reviewed and approved before implementation and be accepted after completion, and supervision after project acceptance is even more complex and difficult. So, a few number of Golden-Sun projects can be supervised in theory, but it is really hard to supervise more than 650 projects at the same time.

As the subsidies for the Golden-Sun projects are allocated beforehand based on application and follow-up project regulation fails to be implemented, it is hard to control the quality of equipment and engineering construction, the scale of actual grid connection and the quality of power stations. It is also difficult to bring the qualities of power generated as the most key part to the applied level, and less than 75% of the projects are put into operation [12]. Huge subsidies offered have not produced ideal results in fueling the demand for China's PV products.

CONCLUSION

The Golden-Sun Program is an emergency policy launched by China to deal with international financial crisis and promote the technological progress and large-scale development of domestic PV power generation industry. The policy measures and supported areas have been adjusted with the changes in the PV market both at home and abroad.

A clearly defined FIT system has far-reaching significance in the shaping of the PV power generation market. Explicit FIT means clear expected returns to investors. In early 2009, China's feed-in policies for renewable energy power had basically taken shape. But the policies regarding PV power FIT have not been made clear, which hinders the development of China's PV power generation market. The Golden-Sun Program was launched as the only choice against the backdrop in which no fixed feed-in benchmark tariffs are available, and an effective means to quickly kick off the market as well.

Under the Golden-Sun Program, subsidies for investment are offered at the initial stage, and huge fiscal funds are used, which has institutional deficiencies. First, the subsidy standards are hard to define with market and cost changes in a reasonable manner; second, the subsidy standards are not assessed on the basis of the quantities of power generated, but on the installed capacity, which plays a limited role in promoting PV power generation; and third, the regulation cost involved in the Golden-Sun Program is very high.

In the Golden-Sun Program, lump-sum subsidies are consumed, but project supervision has not been implemented after governments allocate subsidy funds to enterprises. It is a lesson to be learned in future scaling-up and management of distributed PV power generation projects.

REFERENCES

[1] Zhao Y. Problems in China's PV industry and countermeasures. *Renewable Energy Resources*, 2004 2, 5-8.

[2] Zhao Y. Current development of China's PV industry and prospects. *Solar Energy*, 2012 18, 32-38.

[3] Wang S. Accelerating development of PV power generation to guarantee the sustainable development of China's energy resources. *Engineering Sciences*, 2011 9, 51-63.

[4] Haas R; et al. Efficiency and effectiveness of promotion systems for electricity generation from renewable energy sources: Lessons from EU countries. *Energy*, 2011 36, 2186-2193.

[5] Zhang S; et al. Interactions between renewable energy policy and renewable energy industrial policy: a critical analysis of China's policy approach to renewable energies. *Energy Policy*, 2013 62, 342-353.

[6] Wang Z; et al. (2012). *Report on the development of China's renewable energy industry 2011*. Beijing: China Chemical Industry Press.

[7] Hu Z; et al. Review of wind power tariff policies in China. *Energy Policy*, 2013 53, 41-50.

[8] Zhong-ying Wang., et al.(2013). *Report on the development of China's renewable energy industry 2012*. Beijing: China Chemical Industry Press.

[9] Yi L; Yu Y. Evaluating on the policy system of photovoltaic industry: multi-levels or diversification. *Reform*, 2012 8, 114-124.

[10] Wang C, Wu D. The Golden-Sun program plays an undeniable role in kicking off China's PV market. *Solar Energy*, 2013 16, 18-26.

[11] Zhu Y.(2013). Assessment of the Golden-Sun program. Finance, 5, 72-80.

[12] Li J, Wang S. (2013). *Report on China's PV development 2013*. Beijing: Economy & Management Publishing House.

[13] Shi D. (2012). *Review of energy economics theory and policy research*. Beijing: Economy & Management Publishing House.

[14] Han L. Unsmooth grid connection under the Golden-Sun program [online]. 2013[cited 2013 7 16]. Available from: http://pv.ally.net.cn/special/2013/0716/1589.html

[15] China Institute of Energy Economics Research. (2012). *Compilation of policies regarding renewable energy in China*. Beijing: Economy & Management Publishing House.

[16] PE Morthorst. Policy instruments for regulating the development of wind power in a liberated electricity market [online]. 1999[cited 1999 3 1].Available from: https://www.etde.org/etdeweb/details_open.jsp?osti_id=679597

[17] Judith Lipp. Lessons for effective renewable electricity policy from Denmark, Germany and the United Kingdom. *Energy Policy*, 2007 35, 5481-5495.

[18] Qinghai Science & Technology Department. Golden-Sun demonstration projects implemented in Qinghai produce notable results [online]. 2013[cited 2013 3 5]. Available from: http://www.qhkj.gov.cn/content-96-4903-1.html

The Hydropower Design Institute of China. (2013). *Fruits of construction of China's solar energy power generation 2012*. Beijing: China Water Power Press.

In: Renewable Energy in China
Editors: J. Yuan, G. Luo and J. Chen

ISBN: 978-1-63117-610-4

Chapter 8

EXPLOITATION AND UTILIZATION OF RENEWABLE ENERGY RESOURCES IN RURAL AREAS OF WESTERN CHINA

Guoliang Luo*

School of Economics and Management,
North China Electric Power University, People's Republic of China

ABSTRACT

Western China is confronting two most urgent problems, namely rural poverty and environmental deterioration. Developing renewable energy boasts great potential to solve these two problems. Developing renewable energy in rural areas of western China may bring triple benefits, i.e., meeting rural energy demands, protecting environment and promoting rural development. In 2007, about one million rural households still had no access to electricity in Western China, and traditional biomass energy was still dominant in cooking. Western China boasts plentiful renewable energy sources. Making full use of this advantage will be an effective way to alleviate energy poverty in Western China. Chinese government's renewable energy projects implemented in rural areas of Western China have received notable effect, but realistic barriers still exist in developing renewable energy in Western China. Therefore, strong supports in budget, tax, credit and finance from the government, encouragement to private investment for more financing channels and establishment of a complete rural energy service system are necessary. According to the actual situation of Western china, this chapter brings forward the development mode of the exploitation of renewable energy resources. The Tibet case study finds that the conflict between the need of energy development in the agricultural and pastoral areas of Tibet and its insufficient financial resources will exist for a long time, and this cannot be solved without the support and assistance from the central government and other developed provinces and cities in China.

Keywords: Renewable energy resource, Energy poverty, Western China rural areas

---

* Email: lgl365365@126.com.

1. INTRODUCTION

The past 30 years have witnessed the fastest rate of economic development in the history of the People's Republic of China (PRC). Accounting for 19% of the world's population, the nation of 1.34 billion people has transformed from a primarily agrarian economy to a highly industrialized one.

This rapid change and growth is expected to continue, and with it all the economic, environmental, and social difficulties, contradictions, and problems that other countries have experienced during industrialization. Except, in the PRC, because of the rapid rate of change and the population factor, these changes are occurring in a much more concentrated and intense manner than what has been experienced elsewhere.

The PRC is the world's first largest producer and consumer of energy. While the country as a whole depends largely on coal and oil, nearly 50% of its 200 million rural households in 2008 still relied on burning firewood and various agricultural wastes to heat their homes and cook their meals--a real and serious hazard to the environment and health [1].

In particular, energy security in rural areas, rural environmental challenges, and urban-rural inequality are constraints on the government's ability to achieve its socioeconomic development objectives.

Ultimately, the country must rebalance its economic development so that its rural population, agricultural sector, and environment also benefit from this growth. To make this happen, it was found that the government must invest in rural energy, particularly renewable energy [2].

No developed country has significantly reduced poverty and sustained growth without improving households' access to energy.

Different with traditional energy classification, rural energy mainly refers to the energy consumption and production in rural areas of developing countries. In detail, production energy consumption includes the consumption of agricultural production and rural enterprises, the consumption of living energy include cooking, heating, lighting and household appliances, and the energy consumption of household vehicles. Energy production mainly refers to the production of renewable energy in rural areas.

Rural energy is an important resource and material base for developing rural production, ensuring agricultural living and improving rural environment.

The consumption of rural energy is closely related with regional energy security, soil deterioration, air pollution, rural poverty and change of living modes [3]. So, as for developing countries, how to utilize rural energy and power to relieve energy poverty naturally becomes a focus for energy research.

The chapter consists of seven sections. The second section discusses current situation of renewable energy resources in rural areas of Western China. The third section focuses on energy consumption and its influence in rural areas of western China.

The fourth section analyzes the modes for exploitation and use of renewable energy in rural western China.

The fifth section elaborates the obstacles and actions for developing renewable energy resources in rural western areas. The sixth section is Tibet case study and the conclusion is presented in the seventh section.

2. Renewable Energy Resources in Western China

Western China boasts vast acreage, plentiful energy resources and advantageous renewable energy resources (see Table 8-1) [4, 5]. Every western province has storage of renewable energy resources. But the reserves vary greatly due to different seasons, having some influence to the exploitation of renewable energy.

Table 8-1. Renewable energy resources in rural areas of Western China

Wind energy resource	Solar energy resource	Hydro energy resource	Biomass resource
In Xinjiang, Gansu, Ningxia province and Inner Mongolia Autonomous Region of Western China, the average annual wind power density is 200-300W/m^2, which is even over 500 W/m^2 in some areas.	Western China has rich solar energy resource. The richest solar resource area [kWh/m^2 ≥1750 per year] makes up 42.7% of China's land mass. The second-richest solar resource area [kWh/m^2. 1400-1750 per year] makes up around 17.4 % of China's land mass.	The hydro energy resource of Western China occupies the 81.6 % of countrywide. The total technology development electricity of small hydropower enterprises below 50MW of China is 128MW, of which Western China accounts for 62%.	There are abundant crop stalks, firewood and animal waste in Southwest China, where the weather condition is suitable for developing biogas.

3. Energy Consumption in Rural Areas of Western China and Its Influence

3.1. Characteristics of Energy Consumption for Living in Rural Areas of Western China

Profile of eastern, central and western regions of China is elaborated in Table 8-2[6].

Table 8-2. Profile of eastern, central and western regions of China (2010)

Region	Coverage	Area (/10,000km^2)	Rural Population (10,000)
Eastern	Beijing, Tianjin, Hebei, Liaoning, Shanghai, Jiangsu, Zhejiang, Fujian, Shandong, Guangdong, Hainan	106.8	22708.51
Central	Shanxi, Jilin, Heilongjiang, Anhui, Jiangxi, Henan, Hubei	166.8	23538.12
Western	Inner Mongolia, Guangxi, Chongqing, Sichuan, Guizhou, Yunan, Shaanxi, Gansu, Qinghai, Ningxia, Xinjiang	565.2	22249.17

In 2007, rural China had a population of about 727.5 million, and mostly used traditional biomass energy due to economic and technical restrictions.

Table 8-3. Household energy consumption structures and changes in rural areas of eastern, central and western China (Unit:10, 000 kg ce)

	Eastern China			Central China			Western China		
	Total	Commodity energy	Non-commodity energy	Total	Commodity energy	Non-commodity energy	Total	Commodity energy	Non-commodity energy
2002	12162	3228	8934	11095	2456	8639	13154	2617	10537
	100%	26.5%	73.5%	100%	22.3%	77.9%	100%	19.9%	80.1%
2007	13038	4854	8184	13125	2875	10250	14245	2991	11254
	100%	37.2%	62.8%	100%	21.9%	78.1%	100%	21.0%	79.0%
Growth from 2002 to 2007	7.2%	50.3%	-8.4%	18.3%	17.0%	18.7%	8.3%	14.3%	6.8%

The energy consumption structure in different rural areas varies due to differences in geographical locations, economic and cultural conditions. The structure and difference among eastern, central and western China in living energy consumption indicate: the rural area's living energy consumption in western China is obviously different with that in eastern and central China, which embodies some basic features of the rural area's living energy consumption in western China(see details in Table 8-3) [4,5].

Note: Commodity energy includes coal, electricity, LPG, natural gas, biogas and fuel. Non-commodity energy includes firewood and straws.

Between 2002 and 2007, the non-commodity energy had accounted for about 80% in the living energy consumption in rural areas of western China, higher than that in other regions for the same period.

3.1.1. Energy Consumption Structure with Traditional Biomass Energy as the Dominant Source

The main reasons for the higher consumption and proportion of non-commodity energy in rural western China are the relatively laggard economic development, underdeveloped traffic and high logistics cost in this region. With the continuous increase of family energy consumption demand, rural residents prefer to use free or cheap biomass energy.

According to the category of cooking fuels, the traditional energy use proportion of rural households is the highest in western China, i.e., 66.2%. (See Table 8-4)[7].

Table 8-4. Cooking fuels consumption proportion in different rural areas in 2006 (Unit: %)

Category	National average	Eastern China	Central China	Western Chin
Firewood	60.2	53.1	56.9	66.2
Coal	26.1	18.5	38.4	27.1
Coal gas and natural gas	11.9	27.2	3.8	3.2
Biogas	0.7	0.2	0.7	1.3
Electricity	0.8	1.0	0.2	1.3
Others	0.3	0.0	0.0	0.9

3.1.2. Big Gaps between Western China and Central and Eastern China in Rural Energy Consumption and Service

In 2002, the non-commodity energy in rural living energy consumption in eastern, central and western China were all above 70%, and the difference among these three regions was less than 8%. In 2007, with the rapid rise of commodity energy in rural areas of eastern China, its proportion in total energy consumption surged to 37.2%, 16.2% higher than that in rural areas of western China. The growth rates of rural living energy consumption and commodity energy quantity in central China were both higher than those in western China. In rural living energy consumption, the gap of commodity energy proportion between eastern and western China was apparent. According to the current commodity energy development speed, the commodity energy proportion in rural living energy consumption of central China will keep on rising. The gap between central and western China in commodity energy consumption will be further expanded.

Over a long time, the dual structure resulting from the segmentation of urban and rural areas in China has caused the situation that the rural area drastically lags behind the urban areas in infrastructure construction including electrical facilities. Generally speaking, China's low level of rural electrification is well documented. Great regional gaps exist in rural electricity consumption and grid development. The per capita annual electricity consumption in eastern China is five to six times of that in the central and western regions of China. As for rural annual per capita living electricity consumption, gaps among provinces and municipalities are also quite big. In 2006, the national average per capita living electricity consumption is about 100kWh, while the leading consumption was 434kWh in Beijing, over 14 times higher than that in Tibet (Table 8-5)[7].

Table 8-5. Living electricity consumption in rural areas of Western China in 2006 (Unit: kWh)

Regions	Living electricity consumption per household	Per capita living electricity consumption
Sichuan	358.8	102.86
Guangxi	299.3	71.18
Guizhou	286.4	69.26
Shaanxi	264	66.66
Inner Mongolia	250.7	65.55
Ningxia	267.4	60.31
Xinjiang	217.7	49.26
Gansu	218.6	48.86
Qinghai	223.5	48.54
Yunnan	196	48.2
Tibet	172.3	30.9

Main characteristics of energy poverty in rural areas of Western China are: a number of households in remote mountainous regions have no access to power supply, and most households use traditional biomass energy for cooking.

By the end of 2007, about one million rural households, mainly in western areas (see Table 8-6)[8], had no access to electricity. The rural households without access to electricity distributed mainly in areas served by Sichuan Electric Power Corporation, State Grid Corporation of China and China Southern Power Grid [1].

The areas without access to electricity served by Sichuan Electric Power Corporation distributed mainly in Garze, Daba and Liangshan prefectures and remote mountainous counties (cities, districts), totally 369,900 households in 3,884 villages (1,983 villages no power access at all) of 1,226 townships in 77 counties (cities, districts) of 13 cities (prefectures) [8].

The State Grid Corporation of China totally served 192.308 million rural households of 498,235 villages, of which, 250,000 households and 1,694 villages were unreachable to power supply, or, 99.87% households and 99.66% villages had access to electricity. The China Southern Power Grid totally served 47.614 million rural households of 63,249 villages, of which, 232,800 households and 37 villages (all in Yunnan province) were unreachable to power supply, or, 99.51% households and 99.94% villages had access to electricity.

Table 8-6. Rural households without power access in Western China in 2007

Regions	Rural households (10,000)
Tibet	22.00
Yunnan	23.28
Qinghai	6.16
Sichuan	36.99
Guangxi	2.88
Shaanxi	2.87
Xinjiang	1.03
Inner Mongolia	7.13

3.1.3. Low Proportion of Clean Energy Consumption and Low Energy Efficiency in Western Rural Areas

China has continuously strengthened rural energy construction, and launched lots of projects and funds in boosting rural energy structure adjustment and rural energy consumption mode improvement. Although obvious achievements had been made, China stills lags behind of developed countries in terms of energy consumption modes. In 2010, coal and other regular commodity energy accounted for 29.0%, straw, firewood and other traditional biomass energy for 37.2% in China's rural energy breakdown. However, the quality clean energy of liquefied gas, natural gas and electricity had low proportions [9].

Table 8-7. Per capita living energy consumption changes in rural areas of eastern, central and western China (Unit: kg ce/a)

	Eastern China	Central China	Western China
2002	554	482	621
2007	360	436	506
Decline from 2002 to 2007	-35.1%	-9.8%	-18.5%

According to statistics, the average living energy condition of urban residents is over two times higher than that of the rural residents in Western China. The energy statistics only involves commodity energy. Traditional energy consumption of rural residents is not included in commodity energy since traditional energy is autarkic. Actually, the living energy consumption of rural residents may be higher than that of urban residents. A reasonable explanation to this might be the low energy efficiency of rural energy consumption [10]. The regional changes of rural living energy consumption and the per capita energy consumption changes from 2002 to 2007 show that the per capita energy consumption in rural areas of western China was higher than that in other regions for the same period (See Table 8-7)[5]. This mainly attributes to the overwhelming majority of traditional biomass energy in its total energy consumption. The proportion of living commodity energy of rural western China was apparently lower than that in eastern and central China.

3.2. Consequences of Inefficient Energy Consumption and Services in Rural Western China

3.2.1. Resource and Environmental Problems in Rural Areas of Western China

The low energy use efficiency and outdated energy use mode in rural China cause huge resource loss and environmental pollution. Western China, a region with poorest natural conditions in the country, was trapped by the vicious cycle of "poverty, excessive development, ecological degradation" due to the vulnerable ecological environment, big poverty-stricken population and increasingly severe pollution. Excessive ecological damage has become the biggest obstacle severely restricting the local socioeconomic development.

In the PRC, farm incomes are now under pressure partly because of degradation of their resource base and partly because agricultural wastes from animals and crops and their traditional uses are exacerbating problems in rural environments. Environmental problems include desertification, soil erosion, grassland degradation, salinity on irrigated land, organic matter and fertility loss, burning of crop residues, aquifer depletion, high levels of heavy metals, nitrates and pesticide residues in soils and water, animal wastes, and loss of biodiversity.

Most western rural households continue to rely on directly burning coal or biomass, such as crop stalks and firewood, which are typically burned in low-efficiency stoves for cooking, water heating, and space heating. Most crop straw and fuel wood stoves are 2.5 times less efficient than a biogas or LPG cooker.

Excessive use of crop residues and firewood also exploits forest resources, which has led to significant environmental damage—deforestation, soil erosion, biodiversity reduction, and damage to watersheds. This practice also makes it difficult to recycle chemical elements (e.g., nitrogen, potassium, and phosphorous contained in crop stalks and firewood into the fields, thus reducing the fertility of the soil. In addition, the practice has a heavy social cost, as women and children carry the burden of gathering fuel and preparing the fire for cooking each meal.

Animal wastes from growing livestock farms are constantly introducing massive quantities of non-point source pollutants to waterways. Based on ADB studies, the rural chemical oxygen demand load in 2001 was 1.42 times the load from industry and urban sewage discharges. By 2020, the study projected that the rate of livestock production will have increased by 167% of the 2000 levels. A 40% reduction in rural nonpoint source pollution by 2020 will be needed just to stabilize pollution at its current high levels, which poses a considerable challenge to livestock producers [2].

3.2.2. Inefficient Energy Consumption Leads to Severe Health Loss and Unjustness

Direct combustion of these biomass fuels causes serious air pollution and health risks for the PRC's western rural population. Burning traditional biomass continues to be a major cause of indoor air pollution and respiratory disease in the PRC's western rural areas. Fluoride poisoning is a common health problem in Guizhou province, where some 1.9 million poor farmers are affected, mostly women, children, and the elderly—often from minority ethnic groups—because of inequitable gender roles in the house and inequitable access to efficient stoves or household biogas systems [2].

Each year, about 420,000 Chinese people died from indoor air pollution caused by solid fuel combustion. Respiratory disease has been one of the four major diseases leading to deaths of rural residents for a long time (see Table 8), and accounted for 18% disease-caused deaths in rural area each year, and even seen worse results in Western China [11]. If rural cooking ranges have comparable energy efficiency with urban ones, the mortality caused by respiratory disease will decline 32%. If urban residents consume clean energy instead of traditional biomass energy, the chronic bronchitis of rural children and adults will decrease 80% and 45% respectively [11].

Table 8-8. Proportion and ranking of respiratory disease-caused death of China's rural residents

Year	2000	2001	2002	2003	2004	2005	2006	2007	2008	2009
Proportion in disease-caused death	23.1	22.5	15.6	18.7	13.3	23.5	16.4	17.2	16.9	15.0
Rank	1	1	1	3	3	3	1	3	3	4

Use of traditional fuels for cooking with the attendant pollution and the opportunity cost of gathering them imposes a heavy burden of back breaking and time-consuming job on people particularly women and young girls. Poor households tend to spend more time in collecting these fuels than those from higher-income groups. The need to gather fuels may deprive the young girls from schooling. This 'hard earned' energy is used very inefficiently, converting only about 15% of the total into useful energy [12]. Use of such inefficient and polluting fuels, over time, increases the risks of eye infections and respiratory diseases and it affects the health of women and the girls more adversely, as they spend more time indoors and are primarily responsible for cooking. In the farming and pastoral areas of Tibet, the average time for a woman to collect biomass is 1.5 hours per day, and the time of using the traditional biomass to cook is 2.5 hours per day [13].

4. Modes for Exploitation and Use of Renewable Energy in Rural Western China

Decentralized and centralized developments are the two major modes for renewable energy development in rural western China areas.

4.1. Decentralized Development

Decentralized development is a popular traditional energy development mode in rural China. In comparison with cities, most rural areas were weak in economic base, sparsely populated, laggard in infrastructures, inaccessible to public services and relatively poor in resources. Rural residents often have special firewood rooms or stack firewood and straws directly in their yards. Straws and firewood are the main sources for rural living energy consumption. This traditional decentralized energy consumption complies with the life style and housing condition of rural residents and boasts a lower cost [14]. Renewable energy sources suitable for decentralized development are mainly solar photovoltaics, biogas and straw gasification. Rural areas have inborn conveniences in building solar cookers, solar water heaters and small biogas pits. Solar water heaters can be directly installed on roofs in rural areas, which is simple in installation and low in cost. According to rural households' production and living conditions, decentralized family-size biogas tanks often are built into some micro- circulation systems. Straws and manures of livestock and poultry are materials for producing biogas. Then the biogas produced will be used for cooking, lighting and other living energy consumption. And the biogas slurry and residues can be used to fertilize land or raising pigs, fish and poultries etc. [15].

4.2. Centralized Development

Centralized development mode refers to that the energy production center and users are connected via a transmission network while the energy production and consumption are isolated in location. In this mode, production and supply are independent. Centralized energy production in modern factories are carried out to obtain scale effect and management advantages. Renewable energy users do not need to consider the production and maintenance of energy. Centralized energy development may realize scale economy in energy production and service. However, the construction cost for the energy transmission network is high. And its operation is also limited by the sources and markets of rural renewable energy raw materials.

Different development modes may be applied for the same energy at different areas or at different development stages. Decentralized development mode is applicable to sparsely-populated areas with laggard infrastructures, and centralized development mode to densely-populated renovated rural areas with rapid progress in new rural construction, which are rural areas in eastern China with complete infrastructures, service systems, relatively developed economy and high urbanization level.

4.3 Renewable Energy Projects Implemented in Western Rural Areas

4.3.1. "Light Project"

In September 1996, to respond the initiative of "Light Project" by global non-interconnected areas, Chinese government launched and carried forward the Chinese Light Project, aiming at offering electricity to remote areas by means of rich wind energy and solar energy power supply systems. The National Planning Committee then developed the Action Plan for Phase I of Chinese Light Project in consideration of the West Development strategy, preparing a five-year investment of about 10 billion yuan by national and local budgetary appropriations to solve the power supply problem for about 1.80 million peasant and herdsman households and 2000 administrative villages [16].

4.3.2. "Electricity Access to Villages" Project

In 2002, the Chinese government initiated the "Electricity access to villages" project to build 720-plus off-grid PV power stations in seven western provinces (Tibet, Qinghai, Xinjiang, Sichuan, Inner Mongolia, Gansu and Shaanxi). The Central Government made different appropriations to different provinces, such as,100% appropriation to Tibet, 80% to Qinghai and 20% by the local government, 50% to Xinjiang, Sichuan, Inner Mongolia, Gansu and Shaanxi and 50% by the local governments. The appropriations by the Central Government and the local governments amounted to about 1.8billion yuan. The "electricity access to villages" project has provided 1.3 million people from about 300 thousand households in rural western China with electricity access for basic living, and alleviated the energy poverty in the region [16].

4.3.3. "Golden Sun" Project

Prior to 2008, PV power generation had been mainly used in off-grid projects, such as off-grid power stations in nonelectric areas, household power supply, communication, meteorology, railways and solar energy street lamps. By the end of 2010, the loaded capacity of off-grid PV power generators had reached 16MW, up 20% than the previous year [5].

In order to drive the development of domestic photovoltaic power generation market, Ministry of Finance, National Energy Board and Ministry of Science jointly launched "Demonstrative Building Integrated Solar PV Project" and "Golden Sun" Project in 2009 to support decentralized photovoltaic power generation by lump sum subsidies [8]. "Demonstrative Building Integrated Solar PV Project" and "Golden Sun" Project has provided the great energy to drive large-scale promotion of domestic photovoltaic power generation. By the end of 2012, the projects included in the scope of subsidies had come to 6455MW. Between 2009 to 2012, the subsidy scale provided by the "Demonstrative Building Integrated Solar PV Project" and "Golden Sun" Project are shown in Tables 8-9 [17].

Table 8-9. The subsidy scale of "Demonstrative Building Integrated Solar PV Project" and "Golden Sun Project" during 2009 -2012

Year	2009	2010	2011	2012	Total
Demonstrative Building Integrated Solar PV Project (MW)	90	100	110	225	525
Golden Sun Project (MW)	216	275	692	4747	5930

4.3.4. Replacing Regular Fuels with Electricity from Small Hydropower Stations

Since the reform and opening-up, Chinese government has attached great importance to the role of small hydropower stations in rural social and economic development. The Central Government has allocated special funds raised through national debt issuance to support the local government's development of small hydropower stations each year and carried out the hydropower-electrified counties in five consecutive Five-year plans (1985-2010). The country's policy and fund support greatly inactivate the local governments and people's motivation in developing small hydropower stations. During 1995-2000, the 410 hydropower-electrified counties had completed accumulated investment of 115.1 billion yuan, increased installed capacity of 1,060MW for hydropower, or 3MW increase for each county.

In 2003, in order to solve the energy demands of residents in areas for returning grain plots to forestry, Chinese government carried out the special pilot project of replacing fuel energy with small hydropower in areas with rich small hydropower (5MW and below), high population density, urgent demand to alternative fuels, especially in Western China to protect ecological environment and promote sustainable development of society and economy. The project involved more than 200,000 rural residents in 26 counties of Sichuan, Yunnan, Guizhou, Guangxi and Shanxi, covering 1.56 million mu of natural forest reserves, natural reserves and key soil erosion rehabilitation region with a gross investment of 280 million yuan. In 2006, Chinese government expanded this special project in 80 counties (cities) of 21 provinces (autonomous regions, municipalities), benefiting 636,000 rural residents from 169,000 households with total investment of 1.03 billion yuan. In 2007, China had 1,528 counties equipped with small hydropower stations, including 584 counties with hydropower as the main power supply, and completed a 1.44 billion yuan investment in rural hydropower grid renovation.

The Renewable Energy Law of the PRC enforced in 2006 has profound influence to the development of renewable energy in China. A range of economic incentives like financial subsidy and direct financial investment greatly promoted and bolstered the development of renewable energy, playing a critical role in eliminating rural energy poverty and boosting electrification in rural China.

In April 2010, the Renewable Energy Law of the PRC-Amended stipulated to consolidate the collected electricity price add-ons and national special financial funds into national renewable energy development funds to support the items including renewable energy utilization in rural and pasturing areas and construction of separate renewable energy power systems in remote areas and islands. So, the national renewable energy development funds offer legal and fund supports to the electrification of rural and pasturing areas unfit for grid extension due to restrictions of natural conditions and geographical positions [18].

4.4. Deviations in Exploitation and Utilization of Renewable Energy in Rural Western China

The selection of energy development mode will be a comprehensive consideration of the economic development level, consumption power, residence environment, living habits and infrastructure situation (See Table 8-10) [14]. Infrastructures, a sort of public services, normally need great investment and are restricted by local economic development. In rural areas with developed economy, farmers have sufficient economic power and can afford relatively expensive sources, such as using LPG or electricity instead of firewood in cooking.

In rural areas with high economic development level, infrastructures are more complete and living environment is more suitable for centralized energy development. On the contrary, regions with low economic development are not suitable for centralized energy development concerning living environment and infrastructure development. Decentralized development is the only choice for these regions.

Rural regions in eastern China are relatively developed in economy and society, faster in urbanization, more complete in social infrastructures and much higher in farmer's income in comparison with central and western China. On the contrast, rural western China is underdeveloped, incomplete in infrastructures, low in urbanization and population density, and laggard in living conditions and style, which indicate that rural western China is very suitable for decentralized energy development currently. Practices also prove that most renewable energy projects in western China are developed in a decentralized mode and have received recognition of international communities and organizations.

Table 8-10. Comparison of renewable energy development modes

Items	Decentralized	Centralized
Infrastructures	Poor	Complete
Investment scale	Small	Moderate
Service requirement	Low	High
Market scale	Small	Big
Energy solutions	Solar energy, straw gasification, biomass, small hydropower stations	Wind energy, small hydropower stations, biomass, straw gasification

Recent years, large-scale centralized development has been applied for the exploitation and utilization of renewable energy in western China, such as China's ambitious plan for seven wind power bases. But China's emphasis and support to decentralized development have been insufficient. The fact that there are nearly a million rural households without power access is neglected. Requirements in farmer's living and power demand side in western China are neglected. The rights of local peasants in public service and equitable development are not given sufficient consideration in the development and utilization of renewable energy in western China.

5. Obstacles and Actions for Developing Renewable Energy Resources in Rural Western Areas

5.1. Obstacles for Developing Renewable Energy Resources in Rural Western Areas

Obstacles for developing renewable energy resources in rural western areas are summarized as follows.

First of all, there are no effective or professional electricity regulations. As the regulator of electricity market, State Electricity Regulatory Commission (SERC) regulates the electricity enterprises, electric dispatching transactions and fair and open power grid. The Renewable Energy Act has clearly regulated that power grid enterprises should provide connecting service of electricity generation by renewable energy, and construct and manage the integration system of renewable energy power generation projects. In practice, the power grid enterprises failed to carry out the above regulations, while SERC cannot take effective measures. Due to no effective and professional electricity regulation, the overall grid connection of renewable energy power is hard to be implemented, which leaded to serious consequences. In 2011, wind curtailment percentage of Western China is about 20%, and wind curtailment power is over 7500MWh [19]. This lack of effective and professional electricity regulation leads to waste of renewable energy resources of Western China.

Second, the incentive policies for renewable energy development are not effectively performed. Over a long time, the on-grid price for electricity from small hydropower stations has been low. In 2010, the national average on-grid price for electricity from small hydropower stations was merely 0.28 yuan/kWh, for electricity from thermal power plants was 0.36 yuan/kWh, from nuclear power plants was 0.50 yuan/kWh, from wind power farms 0.55 yuan/kWh, and from solar power plants 1.09 yuan/kWh. The low on-grid price for electricity from small hydropower stations did not reflect the country's policy of encouraging development of renewable energy [13]. In addition, due to the restriction of management systems, the grid for transmitting electricity from small hydropower stations in remote areas of central and western regions has not been incorporated into the nationwide rural grid renovation.

The areas in central and western regions not covered by the rural grid renovation involve 6.82 million people from 1.82 households in 5,420 villages, 904 townships, 184 counties of 19 provinces (autonomous regions) [20].

Most of these areas are former revolutionary base areas, areas inhabited by ethnic groups, remote and border areas and poverty-stricken areas with underdeveloped economy, poor geographical environment and scattered population, which will have high difficulty and cost in rural grid construction and renovation, and high cost for the operation and maintenance of small hydropower enterprises. These factors restrict the sustainable development of small hydropower stations in rural areas of Western China.

Third, financing means are limited. Investment and financing for renewable energy projects in rural China are largely for governmental demonstration projects for public benefit, lacking formal institutional arrangement and special investment and financing systems. Rural areas in Western China have complicated geographical environment, high difficulty in power access (see Table 8-11) [21] and increasing investment risks.

Table 8-11. Costs for power access to non-electric households in rural areas of Western China during 2006-2009

Year	2006	2007	2008	2009
1000 yuan/household	1.01	1.27	1.25	1.42

As for economic and technical feasibility, the development of renewable energy has some advantages in alleviating and solving energy poverty in rural western China. Due to the restrictions of underdeveloped economy, financial straits, slow rural income increase, limited investment modes and financing channels for rural renewable energy construction projects, and lack of improved incentive policies in financing, tax and pricing, a number of challenges in the promotion, transport, installation, training, maintenance and management processes exist in the application of separate renewable energy power generation equipment in rural areas of Western China. This means that releasing and eliminating rural energy poverty will be a tough fight. Limited financing channels and insufficient funds have become the main obstacles for biogas development in China (Table 8-12)[5]. Household biogas development in rural China mainly depends on the Central Government's subsidies and the household's self-raised funds. The local governments offer relatively low funds for biogas development.

Table 8-12. Investment of biogas in China

Year	Total investment (100mil Yuan)	Household biogas projects Qty. Invest. (10,000 (100mil households) yuan)		Small biogas projects Qty. Invest. (Site) (100mil yuan)		Medium and large biogas projects Qty. Invest. (Sites) (100mil yuan)		Service outlets Qty. Invest. (Sites) (100mil Yuan)	
2001	1.3	16.5	1.2	—	—	17	0.1	—	—
2002	3.1	28.6	2.9	—	—	49	0.2	—	—
2003	10.3	103.3	10.2	—	—	24	0.1	—	—
2004	10.3	104.4	10.1	—	—	20	0.2	—	—
2005	10.3	104.6	10.1	—	—	10	0.2	—	—
2006	25.4	261.0	25.0	—	—	44	0.4	—	—
2007	25.4	249.6	24.0	882	0.3	39	0.5	4472	0.8
2008	60.0	419.8	49.0	9271	2.5	187	1.8	40903	6.7
2009	50.0	165.6	23.8	2885	1.6	1579	17.6	18140	7.0

Currently, the household biogas investment in rural areas of Western China is about 5,800 yuan, only 24% of which from the Central Government. The apparently low subsidy to rural biogas projects from the Central Government results in low-quality biogas projects and decreasing utilization ratio, and has a negative influence to its sustainable development.

Fourth, the investment in research and development is insufficient. Insufficient investment in technological development restricts the technological innovation ability of the renewable energy sources in Western China. Compared with the national investment in many demonstrative promotion projects, the investment in the scientific and technological research related to the biogas, wind power, solar energy in the western area in the recent 10 years are obviously insufficient. Many R&D institutions and manufacturing enterprises are reluctant to introduce new products, technologies and materials due to lack of fund. Many key devices for developing renewable energy sources have to be imported. This results in slow development and low industrialization and commercialization.

Fifth, the slow development of socialized services restricts the efficiency of the utilization of renewable energy sources in rural western area. With biogas as an example, in recent two years, China has made a total investment of about 1.4 billion yuan in the construction of rural biogas service stations. However, due to lack of fund for operation, many service stations are under an embarrassing situation that it is hard for them to survive. Due to lack of effective services, the biogas digesters cannot be effectively used. This is a pervasive problem in rural western area, which also exists in the small wind power system and solar photovoltaic power generation system.

5.2. Actions for Developing Renewable Energy Resources in Rural Western Areas

Suggestions to remove obstacles for the development of renewable energy sources in rural western areas:

The first is to establish effective regulation for renewable energies. An effective and professional renewable energy regulation is the important base and premise of developing and utilizing renewable energy resources. First, the function of the National Energy Administration (NEA) for managing the sustainable energy sources industry in a unified way should be strengthened at the same time of establishing effective coordinating mechanism among different departments to clearly define specific responsibilities and execution procedures of related departments and solve the problems in cross authority and omissive responsibilities. Second, a rural and urban renewable energy sources management system should be established to enhance the construction of the rural standard systems for sustainable energy technologies and products, the systems for functional specification, testing and certification and the systems for services related to the energy and the industry, to promote the improvement of the basic conditions for rural residents to use energy and to ensure the safety of rural energy consumption. In addition, for social management of resources and environment related to the development of renewable energy sources in rural areas, explicit regulations should be made. The second is to adopt diversified economically incentive policies and financing mechanism. At the initial stage of the project of renewable energy sources in rural western area, the rural resident may have to make an unaffordable investment.

Therefore, a number of economically incentive policies should be adopted to put up the on-grid price of small hydropower supply, provide subsidies for terminal renewable energy source products and operation of enterprises engaged in manufacturing related products and equipment and make and carry out policies for tax relief. The application of technologies using renewable energy sources brings with huge social and environmental benefits. For this reason, the Chinese government should make compensation for the rural residents. Meanwhile, the rural financial service system should be gradually improved into one serving the rural residents in an all-around way with rural cooperative financial institutions, policy-based financial institutions, postal savings banks and joint-stock financial institutions. The rural financial business should be opened up, and rural residents should be encouraged to fund and develop small hydropower, small wind power and small PV system projects of their own free will and under the support of the government. The third is to expand investment in development technologies related to renewable energy sources in rural western area.

Rural western area has a slow development, with insufficient financial strength. Therefore, the government should lay the emphasis on this area when providing policy support for the construction of new countryside.

During the new round 10-year Western Development, the government should expand its investment in the infrastructure building and development of technologies related to renewable energy sources, promote technologies related to renewable energy sources in rural western areas. So far, China has fully developed a number of technologies and products for using renewable energy sources, such as comprehensive biogas utilization, small hydropower, solar photovoltaic system, small-scale wind power generation technologies. They will be helpful for improving the life and living conditions of the farmers and herdmen in rural western area and finally solving the problems in energy utilization.

In addition, to solve the bottleneck problems in this field, the government should organize the research institutions and innovative new energy enterprises to cooperate with each other to make key breakthroughs by executing key research projects and develop new technologies, new processes, new products and new materials that meet the overall conditions and have prospects. The fourth is to establish renewable energy service system in rural western area. The construction of first-level rural energy service network should be enhanced to solve the problems in promoting the use of renewable energy in the rural area. A number of such service suppliers with integrated functions for technical training, planning and design, purchase of materials, construction and installation, quality control and service after construction should be fostered. Great efforts should be made to put the renewable energy source projects into practice featuring professional construction, property-based management and socialized services.

6. Tibet Case

6.1. Implication of Energy Service in Tibet

6.1.1. Production of Farming and Pastoral Areas

Tibet is called "the third pole of the earth". Because of the unfavorable natural conditions, the various ethnic minority groups there have formed their unique ways of life and production. The fundamental characteristics are as follows: the households mainly focus on the natural economy, maintain the simple reproduction and live a simple life at the mercy of the forces of nature; the farming and pastoral areas of Tibet are sparsely populated, and the farmers are sparsely scattered, with less than 3 people per square kilometer, which makes Tibet the region with the smallest population density; because of the scattered living, the interaction between people is very difficult, and it is very hard to establish and provide public facilities and public services for people's exchanges like other parts in Chinese mainland; in such a relatively closed area, religious belief becomes the important spiritual and living support for farmers and herdsmen.

At present, there is no fundamental change in the mode of production based on the natural economy in the farming and pastoral areas of Tibet. The lack of the necessary conditions for marketization and industrialization in such areas exists, and the biggest challenge is the inadequate supply of energy and other basic conditions and public services.

6.1.2. Necessities for Subsistence

As a matter of course, food and housing are the most basic needs of living for the farmers and herdsmen, and they are the individual needs that can be obtained through market purchase. However, in the farming and pastoral areas of Tibet, meeting the needs of food and housing is inseparable from the energy-related services provided by the government.

Food is the most basic need of human beings. Living in the specific environment, the farmers and herdsmen in Tibet have formed the diet habit of eating beef and mutton as the main food. One important matter related to food is energy supply. At present, cow dung is basically used as the main fuel in the farming and pastoral areas of Tibet and some generating facilities for wind power or solar energy are only established in the places given support and subsidies by public finances.

This kind of energy consumption habit is not only primitive, but also adverse for the protection of the already fragile pastures of the Tibetan Plateau. Without cow dung and other organic fertilizers, the biomass on pastures will be reduced, and then increasing pressures will be given to the already overloaded pastures [22]. Therefore, as for the demand of food, the farming and pastoral areas in Tibet need energy system to support. Generally speaking, energy resources are always viewed and treated as private goods. But in Tibet, because the farming and pastoral areas are sparsely populated, and the purchasing power there is low, it is very hard to solve energy supply through the private supply in market mechanism, and the measures taken by the government are needed.

Because of the cold climate, the heating period in most farming and pastoral areas is more than 9 months in the whole year, and in some places, heating is needed for 12 months. Without a better energy system, the residents will face great difficulty in meeting their basic living standard.

6.1.3. Basis of the Trading of Farm and Stockbreeding Products

The production of Tibetan farmers and herdsmen mainly relies on natural production conditions, so it is hard to lead to many surpluses of farm and animal husbandry products. The transition from natural economy to market economy cannot do without the most fundamental public service systems, such as transportation facilities, information conditions, and financial services. The reality of Tibet shows that these basic conditions are rather inadequate now. Although "To Every Village" project and "Network Program"(access to broadcasting, television, highway, power) project are implemented, the lack of electric power, the message block, and the unsmooth information channel are still the phenomena that restrict the development of the farming and pastoral areas, which cannot be ignored. For example, Qiangtang town, Nagarze County, only over 200 km from Lhasa, but one of the towns with the highest elevation in Tibet, receives newspapers at least 10 days later than in Lhasa. The lack of infrastructure and public products always leads to the failure of exchange by the rural and pastoral residents in Tibet even if they have surplus products. As the infrastructure and public service system which are adaptable to the production and life of farmers and herdsmen are not established, the farmers and herdsmen's expenditure is added, and the rate of capital accumulation is decreased. At the same time, the rural financial institutions in Tibet are few, and the line of credit is low, so the chances are very small for the farmers and herdsmen who want to increase and improve the basic conditions for production and capital goods through financing. As a result, even if the rural and pastoral residents in Tibet can make some income through the exchange of commodities, and then form surplus funds on hand, it is very hard to transform them into productive capital.

6.1.4. Social Implications of Energy Use

Use of traditional fuels for cooking with the attendant pollution and the opportunity cost of gathering them imposes a heavy burden of back-breaking and time-consuming job on people particularly women and young girls. Poor households tend to spend more time in collecting these fuels than those from higher-income groups. The need to gather fuels may deprive the young girls from schooling. This "hard earned" energy is used very inefficiently, converting only about 15% of the total into useful energy [23]. Use of such inefficient and polluting fuels, over time, increases the risks of eye infections and respiratory diseases and it affects the health of women and the girls more adversely, as they spend more time indoors and are primarily responsible for cooking. In the farming and pastoral areas of Tibet, the average time for a woman to collect biomass is 1.5 h, and the time of using the traditional biomass to cook is 2.5 h [12].

6.2. The Status Quo and Features of Rural Energy Consumption in Tibet

6.2.1. Relatively Low Consumption of Conventional Energy Resources

Conventional energy resources mainly include coal, oil products, natural gas, and electric power. Tibet' s per capita consumption of conventional energy is only about 300 kg of standard coal, which is only 1/4 of the national average. In the terminal energy consumption structure, the proportions of the conventional energy are as follows: oil products accounting for 46%, electric power 41%, liquefied petroleum gas 5%, and coal 8%.

Since Tibet is lack of the conventional energy resources, such as coal, oil products, and natural gas, the supply mainly relies on the input from other provinces, which leads to the high cost and the little rural consumption. In 2007, the rural population accounts for about 85% of the total, but the rural electricity consumption only accounts for 3.3% of the gross generation in Tibet, with more than 90% power consumed by the urban residents. The rural areas mainly depend on hydroelectric power, shown in Table 8-13[24].

Table 8-13. Rural electricity consumption in Tibet

Year	2000	2001	2003	2006	2007
Rural electricity consumption/10,000 kWh	3,392	3,117	4,305	6,354	5,553
Electric energy production in Tibet/ 10,000 kWh	66,075	69,690	101,600	151,514	169, 072
Hydropower generation /10,000 kWh	55,350	58,958	92,083	137,901	156,164
Proportion of rural electricity consumption in electric energy production (%)	5.1	4.5	4.2	4.2	3.3
Proportion of hydropower generation in electric energy production (%)	83.8	84.6	90.6	91.0	92.4

Because of the poor living conditions in Tibetan rural areas, the penetration rate of major energy-consuming household appliances is very low, and the electricity consumption in rural areas is low, as well (shown in Table 8-14)[24]. It is not difficult to understand that for the per capita power consumption in rural areas, Tibet is less than 1/3 of the national average, and there is a very large gap between Tibet and other western provinces and autonomous regions (shown in Table 8-15)[7].

Table 8-14. Major durable goods owned by every 100 rural households

Year	2000	2005	2007
Washing machine (set)	2.3	7.0	8.6
Motorcycle (set)	0.2	14.7	33.4
Black and white television (set)	4.8	2.0	1.4
Color television (set)	9.0	47.4	53.9
Refrigerator (set)	0.4	4.4	10.1

6.2.2. Too High Proportion of Biomass Energy Consumption

In Tibet, the fuel-wood energy is mainly consumed by farmers and herdsmen. The people facing a serious shortage of firewood account for about 40% of the total households in the farming and pastoral areas. In most rural areas in Tibet, the main energy resources are dung, firewood and other biomass, supplemented with electric energy and solar power, with the consumption of fuel-wood energy accounting for about 70% in the overall energy consumption. According to a survey, in rural energy consumption, dung accounts for 52.93%, firewood 36.78%, straw 10.17%, and other energy (oil, electric energy, solar power) only 0.12% [13].

In Tibet, the sequence of proportions of major cooking energy resources in every rural household is: firewood, dung, gas (natural gas), solar power, coal, biogas, and electric energy. It is clearly shown in Table 16 that on the energy use for cooking, there is a remarkable gap between the rural households in Tibet and that in western regions and the whole country, and the difference is mainly manifested in the use of coal and dung.

Since Tibet is lack of coal and the transportation cost is high, the use of coal is very limited. Besides, because of the low cost of local rural cattle and sheep manure, and the easy access to them, nearly 50% of the households use them as the cooking energy.

Table 8-15. Electricity consumption of rural residents in China in 2006 (unit: kWh)

Region	Household consumption of every family	Per capita household consumption
Nationwide	388.2	103.2
Tibet	172.3	30.9
Sichuan	358.8	102.9
Guizhou	286.4	69.3
Yunnan	196.0	48.2
Shaanxi	264.0	66.7
Gansu	218.6	48.9
Qinghai	223.5	48.5
Ningxia	267.4	60.3
Xinjiang	217.7	49.3

Table 8-16. Composition of rural residents according to the major use of cooking energy (unit: %) [7]

Region	Firewood	Coal	Gas or natural gas	Biogas	Electric energy	Other energy resources
Tibet	51.9	0.21	1.7	0.05	0.04	46.1
Western regions	56.9	38.4	3.8	0.7	0.2	0.0
Nationwide	60.2	26.1	11.9	0.7	0.8	0.3

6.3. Distribution and Utilization Status of Renewable Energy Resources in Tibet

6.3.1. Hydropower

The exploitable capacity of the hydroelectric resources in Tibet is 0.116 billion kW, and the annual energy production is over 600 billion kWh. The theoretical reserves and the developable resources both make up 30% of China's total, of which the developable resources rank the second in the nation, next only to Sichuan.

The theoretical reserves of Tibet's hydropower resources are mainly in Southeast Tibet, constituting about 60% of all the theoretical reserves of Tibet's hydropower resources. "Sanjiang Rivers" (Jinsha River, Lancang River, and Nujiang River) in eastern Tibet makes up 15% of the theoretical reserves of Tibet's hydropower resources, southern Tibet 24%. The Northern Tibet Plateau accounts for 2/3 of the area in Tibet, but the developable hydropower resources only make up 1%. Yarlung Zangbo River Valley constitutes 87.2% of the theoretical reserves of Tibet's hydropower resources [25].

In Tibet where the conventional energy resources are scarce, the exploration and utilization of hydropower resource plays a significant role in the energy industry of the plateau, and the exploration of Tibet. Because of the restriction of the natural environment, the miniaturization of hydropower generation is the prominent feature. So far, the abundant hydroelectric power resources in Tibet have not been fully developed and utilized.

6.3.2. Solar Energy

Tibet is one of the places in the world with the most extensive deposits of solar energy resource. Table 8-5 shows that there are 1750-3300 h of sunshine annually and there are on average 275–330 days with more than 6 h of sunshine per day in a year.

Table 8-17. Solar energy resource distribution in Tibet [7]

Region	Average annual hours of sunshine (h)	Average annual radiation (MJ.m^{-2})	Area (km^{2})	Average altitude(m)
Lhasa	3,000	7,500-8,000	29520	3,700
Shannan	2,900	7,000-8,000	78,900	3,600
Qamdo	2,180-2,700	6,200-6,500	108,700	3,500
Shigatse	2,600-2,860	700-8,000	417,000	4,000
Nagqu	2,800-2,860	7,500-8,000	420,000	4,500
Nyingchi	1,750-2,010	4,590	117,000	3,000
Ngari	3,300	7,000-8,000	305,000	4,500

The radiant intensity is high, with the total solar radiation of 7000 MJ/m^2. Its geographical distribution increases from east to west, with the most in the west, the second in Dingri on the north slope of Mount Everest, reducing gradually to the southeast, complementary with hydropower. The total amount of solar radiation changes with seasons, with the most in spring and summer, and the least in autumn and winter. However, because of the limited types of solar-related products, the solar energy in Tibet is only used for illumination and boiling water, which is on the low level of utilization, far from satisfying people's energy needs.

6.3.3. Wind Power

The total reserve of wind energy resource in Tibet is about 7730 kW. There are two zones of strong wind in Tibet–one is the area from North Tibet to Ngari, lying roughly on Heihe river road, and the other one is the east section of the valley area between the Himalaya Mountains and Gangdise Mountains. However, because of the topography and the elevation, there is almost no place with the annual average wind speed over 5.0 W/m^2 in Tibet. The areas where the wind power resource can be utilized in Tibet are mainly distributed in the farming and pastoral areas sparsely populated, with the poor transportation services and comparatively backward economy.

It is very difficult to build wind energy farms there, and the operation and maintenance of wind power generators are quite hard even if the farms are established. This kind of natural condition causes the little feasibility of developing large-scale wind generating sets, and it has a brighter future to popularize small-sized wind power generators and utilize them with other resources. Because of the complex terrain in Tibet and the lack of station observation data, the detailed investigation of the actual distribution of wind power and its utility value should be conducted.

6.3.4. Geothermal Energy

The geothermal energy resource in Tibet is characterized by wide distribution, high temperature and abundant deposits. The resources of middle and low temperature are mainly distributed in southern Tibet, western Tibet and northern Tibet. The exploration, development and utilization of geothermal energy resource in Tibet were started in 1970s, and great progresses have been made in geothermal power generation through 30 years' effort. In Tibet, the geothermal energy of middle and low temperature is used for exploration, for example, Yangbajing Geothermal Field, whose generation potential is 34 MW and which can be exploited for 30 years. The installed gross capacity of geothermal power stations is 25.18 MW, accounting for 15% of Lhasa's installed capacity of power network, and it mainly supplies electric power for Central Tibet (Lhasa, Shannan, Shigatse, and Nagqu). After the development of geothermal resources in Yangbajing, Tibet exploited the geothermal fields in Yangyixiang, Nagqu, and Langjiu in succession and built three fairly standard geothermal power stations with the installed gross capacity of 28.18 MW.

Tibet's geothermal application occupies the fires place in China. Nevertheless, since a few matters are ignored in the exploitation of geothermal power fields, such as the scientific research, supervision, environmental protection and sustainable use, the utilization rate of geothermal resources is only 6%.

6.3.5. Biogas

In 1970s, some experts and scholars started to do the exploratory study on biogas in Duiongdeqing County in Lhasa and Shannan. In 2003, Tibet took lead in the experiment on household energy mode of biogas in Shigatse, and the result shows that the application in low temperature has been realized in technology, which changes the traditional mode of biomass energy utilization.

In view of the high altitude, low temperature, and large temperature differences, scientific researchers drew lessons from the mainland experience and technology, successfully built biogas generating pits adaptable to Tibet's climate characteristics, and provided heating and heat preservation for the pits by using the abundant solar energy, to ensure the use of biogas in winter. Up to 2006, 1180 biogas generating pits have been established in Shigatse, Shannan, Nyingchi, Qamdo, and the outskirt of Lhasa, with the average total pit volume of 11800 m^3. The problem of energy use in 2950 people's daily life has been solved, and the ecological engineering system of biogas–solar energy–cultivation–planting–toilets was formed gradually. However, farmers' biogas building lacked related financial support because of the restriction of local economy. It needs 2000 yuan to build a biogas pit of 8 m^3, and some rural households find difficulty in raising funds to build the pit. Most of the farmers do not have the popular science related to biogas generating pits, which leads to the lack of follow-up management on the already built pits; as a result, the comprehensive utilization ratio of biogas is very low. A few rural households even want to give up the use of biogas generating pits.

6.4. Exploitation and Utilization Status of Renewable Energy and Challenges Ahead in Tibet

6.4.1. Utilization of Renewable Energy in Tibet

Since 1980s, the energy construction of Tibetan rural areas has mainly focused on the promotion of solar power, and "Sunshine Plan", "Light of Science", "Electricity Plan in

Ngari", and the project of "making every township have access to electricity" have been implemented in succession. In 2004, over 400 solar photovoltaic power stations were built, with the total capacity of 9000 kW and the total installed capacity ranks first in China. The implementation of these projects plays an important role in the electricity consumption of the farmers and herdsmen without power in Tibet. In 1990s, the "Ride Wind Program" was implemented, various types of wind power generators were successively introduced to Nagqu, Shigatse, Shannan, and Ngari, and other areas in Tibet, and a number of model towns and villages in terms of wind power were established, which make people see the hope and prospect of wind power and solar energy in solving the energy supply in the farming and pastoral areas of Tibet.

There are a huge base number of the rural households without power in Tibet which are widely distributed. The rural power grid is relatively backward and renewable energy resources play an irreplaceable role in rural universal services of energy. In 2005, more than 220,000 households in Tibet were distributed in about 3500 natural villages, and the residential areas without power are widely distributed (see Table 8-18) [8].

Table 8-18. Power-access rate of rural areas in the Tar in 2005 (unit: %)

Region	village	Town	Household
Lhasa	98.2	75.5	86.8
Shannan	96.3	58.3	68.3
Shigatse	55.7	40.7	49.4
Nagqu	77.2	27.8	35.9
Nyingchi	61.8	14.6	64.6
Qamdo	59.4	29.9	43.2
Ngari	94.4	61.3	63.9
Total	71.0	41.0	59.6

Note. The utilization of small hydropower stations and illumination by solar power are included.

Table 8-19. Power supply by isolated small hydropower and solar power in Tibetan rural areas in 2005

Region Town Village Household			
Lhasa	8	24	2,565
Shannan	53	242	14,045
Shigatse	69	421	27,600
Nagqu	83	326	27,373
Nyingchi	28	349	15,803
Qamdo	76	346	33,119
Ngari	33	84	14,690
Total	350	1,792	155,352
Proportion in all the towns, villages, and households respectively	51.2%	30.2%	28.4%

The lighting condition of rural areas in Tibet in the year of 2005 shows that 51.2% of the towns, 30.2% of the villages and 28.4% of the rural households utilized renewable energy resources, such as solar energy, small hydropower and wind power, to solve the electricity problems (see Table 8-19) [8].

Consequently, besides extending the network, it is one of the best ways to make a full use of local renewable resources in light of local conditions for solving the electricity problem of the households without power in Tibet.

On the utilization of solar energy, it is very common for urban residents to use solar water heaters and solar cookers. In comparison, in vast farming and pastoral areas which are relatively sparsely populated, rural households mainly use solar cookers, but the lack of maintenance knowledge causes a high failure rate of the already installed solar water heaters, and the restricted promotion and demonstration. At the end of 2009, there were totally 330,000 solar cookers promoted, one million square meters of passive solar houses, sunshine greenhouses, and solar water heaters, over 100,000 sets of solar photovoltaic household system, and biogas generating pits invested and built for 115,000 households in Tibet.

6.4.2. Main Challenges Facing the Development and Utilization of Renewable Energy in Tibet

To solve the electricity problem of the farmers and herdsmen in the villages without power, Tate Grid Corporation maximized the extension of the network, invested 2.92 billion yuan in total, and solved the electricity problem of 172,000 rural households, investing 17,200 yuan to every household on average. Based on the conditions of the farming and pastoral areas of Tibet, it is very hard on technology and unreasonable on economy to solve the power supply to tens of thousands households without power by extending network. The feasible way is to make a full use of the renewable energy; however, the practical difficulties are following.

(1) High cost of renewable energy's development, utilization and operation

With a low level of urbanization in Tibet, and the sparsely populated farmers and herdsmen, the planning and construction cost of energy infrastructure is very high, and it is difficult to complete by only relying on the local government finance.

The cost of renewable energy's development, utilization and operation is very high. For example, the single capital investment of photovoltaic system or wind/solar complementary system, and the storage battery replacement cost are huge. According to the market price, in power stations used in villages, the construction cost of photovoltaic system is 120,000 yuan/kW, and 95,000 yuan/kW for wind power system; while in the household systems, 95,000 yuan/kW is needed for photovoltaic part, and 50,000 yuan/kW for wind power part. On average, 3-5 years after the station's operation, storage batteries should be replaced, and the replacement cost accounts for about 60% of the total investment Being restricted by the local economy, the rural households lack related financial support on the construction of biogas generating pits. It needs 2000 yuan to build a biogas generating pit of 8 m^3, and collecting the funds for biogas generating pits is difficult to some rural households. Although the technology of developing small hydropower is mature, and the economic cost is relatively low, with the increasing amount of cumulative exploitation, development costs will become a major constraint.

(2) Imperfect energy service system

The energy service system is still imperfect. In Tibet, there is not a mature and corresponding maintenance service network in standard operation. Solar-related products are expensive, and how to ensure that the facilities in every household are of high quality, high reliability and with a reasonable service price are the elements that most users are concerned about. Solar energy industry is investment-intensive and technology-intensive, and there is a shortage of technical and service personnel, which result in the maintenance service backwardness of solar-related products. What is more, since there has not been a scientific testing system, the uneven quality of solar products and the big difference in configuration for products with the same specification create difficulties to farmers and herdsmen in choosing and purchasing solar-related products. At the same time, it is a big challenge to train the sparsely populated residents to properly use the equipment, who have relatively low literacy and weak technological capability. It is of great importance to provide basic training for operating personnel and farmers and herdsmen consuming electricity, and it is an arduous challenge, as well. Most farmers lack the scientific knowledge related to biogas generating pits, and the follow-up management to the already built pits lags behind, which leads to the low rate of multipurpose utilization on biogas. Some rural households even want to give up the use of biogas generating pits.

6.4.3. Realistic Options

Based on the analysis of practice and theory, making a full use of the renewable energy is the feasible way to realize the universalization of access to energy in Tibetan rural households, and the sustainable development.

Nowadays, the population growth, rapid regional economic and social development and the increase in energy demand bring about new challenges for Tibet's ecological protection and conservation of biological diversity. As a result, from the perspective of protecting Tibet's ecological environment and sustainable development, it is imperative for Tibet to implement the strategy of alternatives to fuel-wood energy. The specific methods include: introducing the new technologies and processes of energy development and utilization which are mature at home and abroad, forming a relatively complete system for the large-scale production, selling and energy service, ensuring the wide application and utilization of Tibet's solar energy, hydropower, and biogas, changing the traditional inefficient utilization patterns of biomass energy, gradually increasing the proportion of renewable energy in energy consumption structure, and progressively realizing the transformation from mainly relying on traditional biomass energy to renewable energy-based consumption structure.

The Planning Outline for Alternatives to Fuel-wood Energy in TAR points out that by selecting 3-5 counties of different types in the first batch to carry out the demonstration pilot projects of alternatives to fuel-wood energy, the substitution rate of traditional energy sources comes to 60%.

6.5. Policy Recommendation and Innovation of Investment Mechanism

6.5.1. Policy Recommendation

Governmental responsibilities should be specified, and the mechanism for universal services of energy should be established. The government plays a very important role in universal services of energy. Universal services of energy are one of the important components of public services, and the government should play a leading role in guaranteeing the basic living energy for the public needs. First, it should be stipulated that government departments are the subjects of liabilities to provide universal services of energy. Second, the responsibilities of the central government and local governments for universal services of energy should be specified according to the division of powers and responsibilities. The universal services of energy in Tibet reflect the overall national interests and local interests, so the central government and the government of Tibet should share the responsibilities together. Because of the special financial authority and financial resources of Tibet, at present, the TAR can only raise funds of about 8% in the total expenditure, leaving the largest financing gap among all the less developed provinces. Under such a property rights structure, the major funding should be provided by the central government treasury. According to the Planning Outline for Alternatives to Fuelwood Energy in TAR, the total investment is about 5.811 billion yuan, mainly including the biogas construction in rural areas, the power grid construction, the small hydropower stations, and the project of solar energy popularization. The central government provides more than 80% of the total investment, and the remaining funding is raised by the local government and enterprises.

The mechanism of the central government's transfer payment and partner assistance should be established. The contradiction between the needs of energy development in the farming and pastoral areas of Tibet and its insufficient financial resources will exist for a long time, and it cannot be solved without the support of the central government and developed regions. In the transfer payment, the central government raises the scale of fund allocation. During the"12th Five-Year Plan" period (2011-2015), the annual incremental revenue will be mainly provided for Tibet and other western regions. Besides, for the special transfer payment system for Tibet, the coordination between the central government's special transfer payments and the provincial supports should be strengthened, on the basis of the experience of assistance to various regions of Tibet by mainland provinces and municipalities, coordinated and implemented by the central government. Since 1973, under the guiding principle of helping Tibet to accelerate development and narrow the gap, the central government has implemented a series of preferential policies, offered a large number of financial subsidies, special subsidies and key projects construction investment for Tibet, and organized ministries and developed provinces and municipalities to provide partner assistances for Tibet in human resources, material resources, financial resources, and technologies. For example, in 1994, the central government appointed 13 ministries and commissions, and financial institutions under the State Council, 29 provinces, municipalities, and autonomous regions to jointly provide assistance to the 62 key construction projects needed for the economic and social development of Tibet. These projects cover energy, transportation, telecommunication, agriculture, animal husbandry, forestry, water, grain and oil processing, and social development, with the actual total investment of 4.958 billion yuan, in which the investment of the provinces and municipalities is about 800 million yuan. By the end of 2007, over 30 provinces, municipalities, and central enterprises and over 50 central government ministries and commissions selected and sent altogether more than 3800 cadres to assist and work in Tibet, with the assistance funds for Tibet reaching 9.668 billion yuan [26].

Ecological relocation and poverty alleviation should be combined. Ecological relocation refers to the relocation due to poor ecological environment and the difficulty for human beings to survive. The ecological conditions in many areas of Tibet comply with the standards of ecological relocation. To ensure Tibet's sustainable development under the fragile ecology, the implementation of ecological relocation in some regions of Tibet under adverse ecology is not only consistent with the Scientific Outlook on Development and the requirement of building a harmonious society, but also useful to thoroughly solve the electricity problem for the living and production of farmers and herdsmen in Tibet. There are still farmers and herdsmen in Tibet, living below the absolute poverty line. Tibet has accumulated some experience in poverty alleviation and relocation, which can be combined with ecological relocation.

6.5.2. Innovation of Investment Mechanism

In terms of the investment mechanism, the energy construction in Tibet' s farming and pastoral areas is to achieve two main objectives: one is to solve the use of electricity and energy in the daily life of the farmers and herdsmen without power in order to eliminate energy poverty, achieve the energy fairness, and let the farmers and herdsmen enjoy modern civilization earlier; another one is to provide the energy needed for Tibet's rural economic development and urbanization. To achieve these objectives, policy support and capital investment is crucial. Based on the implementation experience of technology promotion and project construction of renewable energy carried out in Tibet by the government over the years, and based on the feasibility in reality, there are following four options of investment patterns to achieve Tibet's rural development goals of modern energy.

(1) The infrastructure construction mode with all the investment by government, which has been adopted by the already implemented project of "making electricity available to the countryside". Its characteristics are the following. All the construction funds are provided by the government, the implementation speed is fast, and the performance and implementation effects are obvious. However, there must be adequate funds to ensure the complete solution of the power supply for the population without electricity in a short term, and the needed investment by the government is substantial. Besides, the burden of long-term management is heavy after the project construction. There is a need for more specific implementation details on the issues of household systems, product distribution, the definition of property rights, and long-term maintenance services.

(2) The mode of government subsidies plus administrative promotion, in which the government provides subsidies for users through quota or fixed proportions and the enterprises organized or authorized by government at all levels are responsible for implementation. "Brightness Program", local anti-poverty projects, and the construction of electrification counties which are already implemented have all adopted this approach. This method is relatively robust and quick, but it is hard to achieve the optimization of resource allocation.

(3) The market promoting mode under government guidance, which is similar to the mode adopted by the China Renewable Energy Development Project (REDP) implemented by the National Development and Reform Commission, the World Bank, and Global Environment Facility. By providing merchandising allowances for enterprises instead of the direct subsidies to users, establishing the grants for technical progresses, setting up market development fund, and other means, this project gives support to the renewable energy companies' operating activities of marketization and commercialization, supports institutional capacity building, establishes consumer protection mechanism, creates a market environment,

encourages fair market competition, helps enterprises to gradually reduce production costs, raises product quality, increases marketing sales, and improves after-sale services. By providing sustainable energy and power supply for users in remote areas through market operation step by step, the funds burden for the government will become light, and the development will be more sustainable. But because of the restriction of users' purchasing power and other elements, it will take a long time to achieve the electricity goal, and the market barriers need to be eliminated gradually.

(4) The market operating mode under government subsidies and supervision. According to the current situation, experience and development direction of China's renewable energy market, the market mechanism should be made to play a greater role as far as possible by making use of the current advantages. Enterprises should be encouraged to invest in the energy construction of Tibet, and the government should offer subsidies and support to the enterprises through preferential policies and incentives. For example, the mechanisms similar to "Wind Concession" can promote the sustainable development in renewable energy industry.

CONCLUSION

Relatively backward infrastructures and public service systems in rural areas of Western China restrict the development of the local economy and the income increase of local farmers and herdsmen. In the energy consumption of this area, the traditional biomass energy takes a higher proportion, while the clean energy application is insufficient. Most of China's households without access to electricity are distributed in the western remote mountainous areas of China. Therefore, it still has a long way to go to relieve or eliminate the energy poverty in this area.

The rural western area has abundant renewable energy sources. To develop and use these energy sources is an effective way to relieve or eliminate the energy poverty, increase farmer's and herdsman's employment opportunity and income.

In general, in the rural western area, establishing effective regulation for renewable energies is the important base and premise of developing and utilizing renewable energy resources. Secondly, it is needed urgently to adopt diversified economically incentive policies and financing mechanism. Thirdly, investment in development technologies related to renewable energy sources must be expanded. Finally, a renewable energy service system must be established.

Based on the implementation experience of technology promotion and project construction of renewable energy carried out in rural western area (including Tibet) by the government over the years, and based on the feasibility in reality, there are following four options of investment patterns to achieve Tibet's rural development goals of modern energy.

REFERENCES

[1] Bai J. Great challenges during the 11th five-year plan period [online]. 2012[cited 2012]. Available from: http://studentweb.hhs.se/courseweb/CourseWeb/Public/9998/0803/Water%20pollution%20in%20 China.pdf

[2] Zhang, Q; et al. (2010). *Rural biomass energy 2020.* Mandaluyong City, Philippines: Asian Development Bank.

[3] Tonooka Y; Liu J; Kondou Y. A survey on energy consumption in rural households in the fringes of Xi'an city. *Energy and Buildings,* 2006 11,1335－42.

[4] Yang M. (2013).*China's New Energy.* Beijing: China Water& Power Press.

[5] Wang Z; Ren D; Gao H; et al. (2013). *The Renewable Energy Industrial Development Report 2012.* Beijing: China Economic Publishing House.

[6] National Bureau of Statistics of China. (2012). *China Statistical Yearbook2011.* Beijing: China Stats press.

[7] National Bureau of Statistics of China. (2008). *The 2nd National Agricultural General Survey of the PRC.* Beijing: China Stats press.

[8] State Electricity Regulatory Commission. Investigation into China's rural households without access to electricity [online]. 2009[cited 2009 5 7]. Available from: http://info.electric.hc360.com/2009/05/07084583027-4.shtml

[9] Du Y. Developing clean energy and optimizing rural energy structure. *Energy*, 2011 1, 1-3.

[10] Shi D. Study on regional difference in energy efficiency in China and related reasons. *Management World*, 2008 2, 27-32.

[11] Li K; Liu C; Wei Y. Analysis on China's current energy poverty. *Energy of China*, 2011 8,1-9.

[12] Reddya B; Balachandrab P; Hippu S. Universalization of access to modern energy services in Indian households-economic and policy analysis. *Energy Policy*, 2009 11, 4645-4657.

[13] Feng T; Cheng S; Min Q; Sun W. Biogas energy model and benefit analysis in Shigatse prefecture of Tibet. *Resource Science*, 2008 9, 13-19.

[14] Liu X. Comparative analysis of urban and rural biogas development modes. China Population, *Resources and Environment*, 2010 3, 302-305.

[15] Hu Q. Analysis on development models for the utilization of rural biomass energy. *Biomass Chemical Engineering*, 2008 3, 26-30.

[16] Luo G; Zhang X. Universalization of access to modern energy services in Tibetan rural households-Renewable energy's exploitation, utilization, and policy analysis. *Renewable and Sustainable Energy Reviews*, 2012 6, 2373-2380.

[17] Meng X. Decentralized Photovoltaic Power Generation of China. *Solar Energy*, 2013 7, 4-12.

[18] Ren D. New policy saved new energy? *Energy Review*, 2010 6, 114-120.

[19] State Electricity Regulatory Commission. The Report on Wind Power Accommodation Regulation [online]. 2012[cited 2012 8 3]. Available from: http://news.bjx.com.cn/html/20120803/377820.shtml

[20] Zhao Z. Open the door to investment in small hydropower. *Energy*, 2012 3, 36-42.

[21] Luo G, Guo Y. Rural Electrification in China: A policy and institutional analysis. *Renewable and Sustainable Energy Reviews,* 2013 23, 320-329.

[22] Gang L; Mario L. Rural household energy consumption and its impacts on eco-environment in Tibet. *Renewable and Sustainable Energy Reviews*, 2008 7, 1890-1908.

[23] Qiang W; Huan N. Situation and outlook of solar energy utilization in Tibet, China. Renewable and Sustainable Energy Reviews, 2009 2, 2181-2186.

[24] Tibet Bureau of Statistics. (2010). *Tibet statistical yearbook-2010.* Beijing: China Statistics Press.

[25] Li W. Problems of Tibetan rural sustainable development and countermeasures. *Energy Research and Information*, 2008 4, 193-198.

[26] Dong S. Partner Assistance to Tibet, problems & countermeasures. *Journal of Guangdong Technical Vocational College*, 2009 6, 20-24.

In: Renewable Energy in China
Editors: J. Yuan, G. Luo and J. Chen
ISBN: 978-1-63117-610-4

Chapter 9

TRANSITION TO RENEWABLE-BASED ENERGY SYSTEM IN CHINA

Jiahai Yuan
School of Economics and Management,
North China Electric Power University, People's Republic of China

ABSTRACT

To achieve higher renewable energy (RE) shares in China will require additional integration efforts starting now and continuing over the longer term. These include improved understanding of the RE resource characteristics and availability, identification of possible integration pathways, investments in enabling infrastructure and research, development and demonstrations (RD&D), modifications to institutional and governance frameworks, capacity building and policy package in anticipation of RE growth. As a concluding chapter of this book, this chapter addresses the following topics: appraising the potential of various options integrating various RE sources into energy supply and final sector consumption in China, identifying the possible transition pathways, analyzing the main barriers to RE integration, and proposing policy suggestions to deploy energy system transition in China.

Keywords: Renewable energy system; transition pathways; facilitating policy; China

1. INTRODUCTION

Though energy system transition has been extensively studied recently, the magnitude and complexity of the challenge has not yet been widely recognized. As noted by Larson (2009), "*The transformation of existing energy systems to sustainable and renewable alternatives represents a tremendous management challenge. We need to view it as an opportunity to renew energy systems and renew important parts of the economy and social life...So far, primarily the technical aspects of different alternatives have been debated. We know a lot about 'what' has to change. So far, however, very little has been written about the 'how' of the change process*" [1].

Established energy supply systems are relatively new in terms of Chinese civilization history, with100 years for the coal supply system; only around 60 years of national grid electricity; and 50-60 years for oil and gas supply. Based upon the rate of development of these historical precedents, under enabling conditions and with societal acceptance, RE systems could conceivably become more prominent components of the national energy supply mix within the next few decades. Energy systems are continuously evolving, with the aims of improving conversion technology efficiencies, reducing losses, and lowering the cost of energy services to end users. On the one hand, as part of this evolution, it is technically feasible to continue to increase the shares of RE through integration with existing supply systems at national, regional and local scales, especially along with the maturity of RE technology and improvement in the economic efficiency. On the other hand, current energy system, which is carbon-locked in China, is strongly resistant to change and may hinder the penetration of RE. To enable RE systems to provide a greater share of heating, cooling, transport fuels and electricity may require modification of current policies, markets and existing energy supply systems over time so that they can accommodate greater RE at higher rates of deployment.

2. Sustainable Energy Development Pathways in China

2.1. Renewable Energy Options

Generally speaking, RE can be integrated into existing and future centralized energy supply system, via power grids, oil & gas pipeline, heating and cooling networks, etc. on the opposite, RE can be directly deployed in the end-use sectors, including transport, industry, buildings and agriculture. On both sides, there are two kinds of different solutions to RE, integration into centralized, high voltage power systems, district heating systems, and liquid fuel and gas pipelines, or, integration into distributed, small-scale energy systems such as low voltage power systems, heating and cooling of individual buildings, and liquid or gaseous fuel production for local transport use.

Numerous studies have indicated that the potential of integrating RE into current power systems is tremendous [2-3]. This argument is especially true for China, because most of the RE resources, including hydropower, wind power, solar power, biomass and ocean energy, can solely or mainly be utilized by way of power generation. Also, most of these RE resources (hydropower and wind power in particular) are located in remote area and far away from the demand, and under this situation, the most convenient way of integrating RE into the energy supply systems is through power systems. Meanwhile, there are regional heating systems in Northern, North-western and North-eastern China (or regions located northern of the Yellow River) which are currently fueled by coal combustion. Retrofitting existing heating systems in combination with local RE resources input, including solar power, geothermal energy or biomass energy can also substantially cut the fossil fuel consumption. Or, with the installation of heat storage equipment, renewable power as solar and wind can directly be utilized for regional heating. Geothermal energy can be directly used for heating purpose, or can be used for power generation, depending upon the temperature of the extracted underground fluid. Biomass energy includes a variety of resources and correspondingly there are a variety ways of integration. Most of the raw biogas produced from landfills, urban sewage, industrial and agricultural wastes is used onsite or distributed in dedicated local gas systems, primarily for heating purposes. Therefore, there is no extra requirement on the integration of biogas. For integrating biofuels into the liquid fuel systems, the existing petroleum-based infrastructure can be used, with some adaptations. For ocean energy, it can only be utilized by power generation.

The options of deploying RE in end-use sectors vary. For example, in transport sector, energy efficiency improvement in traditional vehicles, biofuels from biomass crop as sugarcane or cotton, hybrid electric vehicles (HEV), and plug-in electric vehicles are possible options. For China as the most populous developing country with limited arable land, biofuels may not be a feasible option. Electric vehicles are feasible options but its deployment is dependent on the maturity of EV technology and the charging infrastructure. In buildings and households, decarbonization can result from integration of RE in electric power systems, heating and cooling networks and gas grids, or by installing onsite RE technologies directly integrated with the building structure. Energy efficiency measures can be combined with RE deployment in a building. For example, in China, the deployment of solar water heater is very successful. A total of 168 million square meter solar water heating system has been installed in urban and rural households [4]. Roof solar PV, passive solar buildings and heat pump also have the similar potential but their deployment is subject to the impact of multiple factors including technology, price, standards and regulations on buildings and power grids. Industry accounts for 70% of primary energy consumption in China. The sector itself in highly diverse, ranging from "heavy" energy-intensive basic material manufacturing (including iron and steel, non-ferrous metals, chemicals, minerals, and pulp and paper industries) to "light" manufacturing. The integration of RE in industry is diverse, sector-specific and process-specific, including direct use of process residues and biomass-derived fuels for on-site heat and CHP (for example, in pulp and paper industry), direct recovery of residual heat for on-site heat and CHP (for example, in iron and steel industry), and direct use of RE (solar thermal or geothermal energy) for process heat and steam demands, etc.

2.2. Other Options

There is a long way for RE to develop into primary energy sources from the current "alternative ones". The implication is that, before RE can contribute a significant share of energy service in China, other options must be secured to ensure energy security and supply adequacy in China. Based on comprehensive appraisal of resource endowment and deployment potential, the followings are feasible options in China's future energy transition pathways:

- **Energy efficiency**

Energy efficiency is the best energy option in that it can directly cut energy demand, and reduce the required capacity and cost of the energy supply system. For China, though remarkable progress has be attained during the past decades, especially during the last Five-year-plan period (2006-2010), there is still vast potential for future improvement considering the big gap between China and its developed counterparts [5]. Energy efficiency improvement can come from technical progress in thermal power generation and heavy industries, more efficient buildings and appliances, efficiency improvement in transport, etc.

- **Nuclear power**

Nuclear power is not a renewable energy, but is a clean generation technology without GHG emissions. Unlike in France, the United States and other developed nations where nuclear holds major portion in the generation mix, historically the share of nuclear in China is trivial. However, there is a strong sentiment against nuclear power worldwide, especially after the serious nuclear accident in Japan in 2011 [6]. In China, all the under-construction nuclear power projects were suspended and no new projects were approved after the accident. With extensive safety appraisement and comprehensive discussion, China lifted the ban in the end of 2012 and now China has the biggest scale of nuclear power under construction in the world. China is also one of the major and active players on the next generation nuclear technology. In the coming decade or more, when hydropower reach its upper limit of technically and economically exploitable resource, and renewable sources as solar and others can hold a marginal position in the overall energy mix, nuclear power will play an irreplaceable role in China low-carbon energy development.

- **Natural gas**

Natural gas is fossil energy, but is much cleaner relative to coal in term of emission factor. Natural gas is a principal source of energy in Europe and the United States. But in China, due to insufficient resource reserve, the role of gas in China's energy mix is trivial until recently. However, the exploration of gas reserves (including nonconventional coal bed methane and shale gas)in recent years has made encouraging progress and may significantly expand the number of exploitable years [7], which provides the possibility for gas power development in China. Also, China has secured several big international contracts with gas exporter countries as Australia, Russia and Turkmenistan, implying that there will be plenty of gas supply in China in the coming years. The substitution of coal by gas will certainly ease the burden of GHG emissions pressure in China.

- **"Clean" coal**

The role of carbon capture and storage (CCS) in global energy transition is still controversial. The strongest opposition is that CCS doesn't make sense because it can't eliminate but just capture and storage CO_2. Besides, though the main components of CCS are available in different sectors, currently there is no commercially available integration system. Therefore, the large-scale deployment of CCS will inevitably hinder the development of other available options such as wind power and solar power, as well as the future possible technology as hydrogen. Though there are many novel technologies under laboratory research and it is a must to shift the future energy system from currently carbon-based to decarbonized, they are still in the very early stage of technology development. The experience with the development of major energy technology indicates that it takes at least 40-50 years to mature a novel technology [8].

Table 9-1. Energy technology options analysis

Stage of (technology) option	Fossil-fuel based	Nuclear	Renewables	Demand-side technologies	Energy carriers and storage technologies
Mature	USC boilers Combined heat and power generation (CHP) Gas (coal bed gas) power generation	Existing fission reactors	Wind turbines Traditional geothermal technology solar heat water system Biomass generation	Energy-efficient appliances Compact fluorescent light Industrial process innovation House insulation	Batteries Pump storage
Early commercialization (6-15 years)	Some gasification technologies Some CO_2 capture technologies CO_2 storage Retrofitting of old boilers to USC	New fission reactors	Hydropower Some wind turbines Biomass boilers Biofuels PV and concentrated solar power (CSP)	Heat pump (geothermal or air) Passive solar Low-carbon buildings LED lighting system	Fuel cells
Development and demonstration (D&D) stage (15-20 years)	Some CCS technologies Integrated gasification combined cycle (IGCC) Underground coal gasification Polygeneration Fischer–Tropsch process	Fourth generation reactors (high temperature gas reactors etc.)	Grid modification Wave Tidal Biofuels, e.g., gasification, AD, transport fuels Grid modification	Intelligent buildings Smart meter The internet of Things	Hydrogen from gas and electrolysis Fuel cells
Research stage (20 years and beyond)	Novel CO_2 capture technologies	Nuclear fusion	Biofuels e.g., pyrolysis Enhanced geothermal system		Hydrogen generation from biomass, waste, nuclear, etc.

On the other hand, because of the limit of resource availability and the technology characteristics, currently matured renewable energy technologies like wind and solar are not enough to meet with the increasing energy demand. Therefore the unavoidable though unfavorable reality is that fossil energy will still dominate the energy mix at least until the middle of the century until proven next generation energy technologies scale up.

For China, even with great efforts on renewable energy development, coal will probably hold some 50% share in the total primary energy supply. Therefore, in the long run, CCS is the only technology available to mitigate greenhouse gas (GHG) emissions from large-scale fossil fuel usage [9]. Especially for China, in an even stricter global climate change policy environment, "clean" coal technology as CCS has strategic importance. In the short run, more efficient thermal power generation technologies as CHP and USC can help improve energy efficiency and reduce the GHG emissions in power generation.

A detailed analysis on the possible technology options for energy system transition in China is presented in table 9-1.

2.3. Overall Transition Pathways

In system transition and transition management literature, different authors have proposed different topologies for transition pathways depending on whether they believe that transition is manageable or not [10-13]. In this chapter, we adopt a typology of transition in the school of innovation transition as, reproduction: on-going processes of change within the socio-technical regime (i.e., not involving interaction with the landscape or a technological niche); transformation: processes of change that arise from the interaction of an evolving landscape with the socio-technical regime (but not with the technological niche level); substitution: replacement of one dominant technology within the socio-technical regime by another as a consequence of interaction between all three levels; de-alignment/re-alignment: interaction between the three levels resulting in competition between a dominant technology within the regime and a number of other competing options, which have different performance characteristics, eventually resolved through emergence of a new dominant option; and re-configuration: replacement of a set of interlocking technologies by an alternative array of inter-related technologies which fulfill the same, or similar functions [13].

To break the carbon lock-in, the feasible options for China are: 1) a more balanced economic growth plan, which calls for an effective and practical development policy, especially a policy for industry; 2) more energetic efforts on energy efficiency by technology innovation in manufacturing sector, improved energy efficiency in buildings, high energy efficient appliances, etc., especially in household and transportation sector, a comprehensive city, transportation and building planning would be vital to avoid the potential future demand; 3) more penetration of clean energy sources in primary energy supply, especially more radical development of hydro, wind and solar power in the near future, 4) more joint research & development and commercialization efforts with international partners to deliver CCS and carbon gasification technology in the medium future, considering the unavoidable role of coal in China; 5) more active science and technology research input in fusion nuclear and next generation hydrogen alternatives in the hope to replace the current hydrocarbon based energy sources. In exploring the potential transitions of China's energy system from a long-term perspective, we are interested to probe the overall canvas and will not focus on technique or engineering details. Because of the endogenous uncertainty in system innovation processes and the vast complexity of the topic itself, the picture is destined to be imperfect and speculative. A two-stage approach is employed in the study, with the first stage covering 2010-2020 while the second covering 2020-2030 and beyond. Then different technological options are classified into reproduction, transformation, substitution, reconfiguration and de-alignment/re-alignment pathways according to their roles in the transition process. Table 9-2 lists the possible options in every pathway for the two stages. For the first stage, the feasible pathways will centre on reproduction, transformation and substitution. In the reproduction pathway, considering the already high efficiency of coal power plants, the effect of regular efficiency enhancements in existing units would not be significant, but replacement of small-scale inefficient units with UVC or CHP units is promising. In the transformation pathway, technology innovation in industrial process, popularization of energy efficient appliances, refurbishment of existing housing to increase energy efficiency, promotion of distributed generation in remote regions with alternative clean energy sources would be priorities. In the substitution pathway, developing safer fission nuclear power in eastern coastal regions, developing large-scale inland and onshore wind power bases, CHP gas power in relatively developed eastern regions, small-and-medium scale waste (and biogas) CHP generation to provide local grid would be priorities. Particularly, solar power could be promoted by employing stronger policy incentives to breed other important renewable alternativesin China as quickly as possible. Heat pump and passive solar technology which can substitute power demand substantially in household and office buildings could also be promoted. During the first stage, though there will be no obvious progress in reconfiguration and de-alignment/re-alignment pathways, proper R&D input in related technologies, especially CCS, coal gasification, hydrogen production, fuel cell, next generation nuclear power etc. and investment in their demonstration & commercialization would be important for the second stage. For such a tremendous task, hand-in-hand cooperation with international partners on selective promising technologies is a priority. For the second stage, with the increasing share of RE in different scales and breakthroughs in next generation nuclear and energy carrier/storage technologies, a multi-layer power system, consisting of large centralized units such as UVC + large scale hydro + nuclear + gas units interconnected by an ultra-high voltage DC/AC transmission system as the upper layer (Macrogrid), thousands of regionalized medium scale RE, CHP gas and/or biogas with fuel cells and other power storage facilities as backup interconnected to Macrogrid as the middle layer (Mesogrid) and millions of localized small-and-medium renewable, MicroCHP with small scale fuel cells, interconnected with the regionalized Grid as the bottom layer (microgrid) could be gradually shaped.

3. Integrating Renewable Energy in China

3.1. Integration Requirement

The rapid growth of electricity demand in China, coupled with the geographically dispersed nature of many renewable sources (see chapter one) makes electricity an attractive energy vector to harness Re where adequate network infrastructure is or can be available.

Table 9-2. Pathways for energy system transition in China

Pathway \ Stage	First stage: 2010-2020	Second stage: 2020-2030 (and beyond)
Reproduction: efficiency enhancement in coal-based large-scale power system	Regular efficiency enhancement; CHP; substitution of small scale inefficient unit with UVC unit; power grid operation optimization	Same as first stage
Transformation: minor modifications to coal-based large-scale power system	Demand side options as technology innovation in industrial process, popularization of energy efficient appliances, house refurbishment such as insulation and double glazing ; isolated distributed generation (DG) in remote areas	Low carbon building
Substitution: technology substitution or power demand substitution within centralized power system	Gas power, waste (biogas) power and/or gas + CHP; fission nuclear power; inland and onshore wind power; solar power in demonstration and commercialization; heat pump; passive solar	Gas + CHP power; GIF nuclear; offshore wind power; advanced solar power; fuel cells; MicroCHP
Reconfiguration: major modification to coal-based large-scale power system	CCS research &experiment & demonstration; Fuel cells research & experiment & demonstration; Centralized power grid+ solar/wind/hydro/biogas and other combined DG	Regional power grid + ultra voltage transmission connection + expansion of kinds of DG technologies including fuel cells
De-alignment/re-alignment: power system coevolves with hydrogen/ fuel cell technologies	Research & Development of hydrogen technology	Possible Macrogird with large-scale UVC+CCS units and other traditional carbon-free units or large scale fuel cell plants based on underground coal gasification +hydrogen production as carriers +Possible DG and Microgrid with fuel cell, wind, solar and other units

With the development of EV and heat pumps, electricity is also taking a growing share in the transport and heat markets. Additionally, with the development of effective communications systems and technologies as well as smart meters, power system is experiencing dramatic change in China. All these potential developments, RE, demand side participation, EV and any new traditional power generation, need to be integrated into power systems. They collectively and individually pose common and unique challenges.

RE resources as hydropower, wind power and solar power are distributed far away from load center in China. Therefore, renewable-generated electricity needs to be transported over considerable distances (ranging between 500 and 1500 Km). In this situation, additional transmission infrastructure may be needed to enable access to higher quality RE resource regions by electricity load centers rather than utilizing lower quality RE sources closer to load centers. Certainly, some RE resources can also be exploited as embedded generation in distribution networks, which may result in benefits for the system when at moderate penetration levels, but also can pose challenges at higher penetration levels.

Besides transmission and stability challenges, RE resources as wind and solar power lack the flexibility needed to deal with real-time balancing requirement of power system operation. This is because they are subject to variability across a wide range of time scales important to power systems and also experience more uncertainty in predicted output. Furthermore, renewable plants may displace non-renewable plants that have heretofore provided the required flexibility. Some renewable sources (hydropower with reservoirs and bio-energy) may partly contribute to provide flexibility. However, overall balancing will become more difficult to achieve as partially dispatchable RE penetrations increase. The serious wind curtailment in North-eastern and Inner Mongolian power grids is a typical case exemplifying this issue [14].

There are also challenges with regard to very short-term system balancing (i.e., frequency response). At high penetration levels the need for frequency response will increase unless supplementary controls are added. Many of the renewable technologies do not lend themselves to such service provision. In addition, RE interfaced through power electronics may displace synchronous generators, thereby reducing the overall system inertia and making frequency control more difficult. In China Research& Development activities are in progress to deliver with frequency response from time variable sources such as wind turbines, and some equipment with frequency response and inertial response is already available.

Besides system operation challenges, there is additional challenge of managing the transition from the coal power dominant generation mix of today to future renewable-based low carbon power systems in China. Major changes will be required to the generation plant mix, the system infrastructure and operational procedures if such a transition is to be made. Specifically, major investments will be needed and will need to be undertaken in advance, so as not to jeopardize the reliability and security of power system. Recall that planning economy is still the standard procedure in China's power sector and in China the government is a developmentalism one, even when the technologies are mature enough to solve the security and reliability issues arisen from higher share of RE integration, RE friendly operational procedures may be difficult to be established. Table 9-3 provides an overview on the integration characteristics of different RE technologies [15].

3.2. Integration Options and Measures

Though there are multiple challenges for integrating high share of RE in power system, fortunately the general form of the solutions is already known today.

Table 9-3. Summary of integration characteristics for selecting renewable energy technologies

Technology		Plant size (MW)	Variability (time scale)	Dispatchability	Location diversity	Predict-ability	Capacity factor (%)	Capacity credit (%)	Active power, frequency control	Voltage reactive power control
Bio-energy		0.1-100	seasons	dispatchable	moderately	high	50-90	similar to CHP	full	full
Solar power	PV	0.01-100 modular	Minutes to years	hardly	high	low	12-27	<25-75	good	good
	CSP	50-250	Hours to years	partially dispatchable	moderately	high	35-42	90	good	good
Geothermal energy		2-100	years	dispatchable	N/A	high	60-90	Similar to thermal	full	full
hydropower	Run-of-river	0.1-500	Hours to years	partially dispatchable	moderately	High	20-95	0-90	full	full
	reservoir	1-20000	Days to years	dispatchable	moderately	high	30-60	Similar to thermal	full	full
Ocean (tidal and wave)		0.1-300	Hours to days	hardly	moderately	high	20-60	10-20	good	good
Wind power		5-300	Minutes to years	hardly	moderately	low	20-40 onshore 30-45 offshore	5-40	good	full

There is already considerable experience operating power systems with large amounts of renewable sources, and integration studies have also offered valuable insights. Here in this section, some general forms of options that can be used in China will be outlined.

- **Improving network infrastructure**

Strengthening connections within an electrical power system, or introducing additional interconnections to other systems, can directly mitigate the impact of variable and uncertain RE sources. With strengthened connections, electric power can be more easily transmitted from where it is generated to where it is consumed, without being constrained by bottlenecks or operational concerns. This argument is especially true for China, where RE resource distribution and load demand is not matched geographically. Operating as part of a larger balancing area, or sharing balancing requirements across regional power systems, can reduce the integration costs associated with RE generation and the technical and operational challenges. Opportunity exists within or across region in China's power system by aggregating different RE resources. For example, the concept of bringing together wind power and solar power in northern China can be feasible in that their output hours are complementary in a day.

In China, there is transmissions system expansion plan along with the development of ten wind power bases at 10 GW scale. Especially, the ultra high-voltage transmissions (UHV) lines are under construction to enhance the trans-regional transmission capacity of RE generation [16]. In China, UHV is an important part of the concept of Super Smart Grid, which was proposed by China's major power grid operator, SGCC (see chapter 6). Delivering new network infrastructure will face institutional challenges, in particular to provide incentives for the required transmission investments. However, such an incentive is at least problematic or unclear for grid operators in China under existing pricing mechanisms because there is no separate transmission & distribution tariff in place.

- **Increased generation flexibility**

Thermal generation provides most of a power system's existing flexibility to cope with variability and uncertainty, through its collective ability to ramp up, turn down and cycle as needed. Great flexibility is required from the generation mix for higher penetration of variable RE generation. This can imply either investment in new flexible generation or improvements to existing power plants to enable them to operate in a more flexible manner. Thermal power plants can be designed or retrofitted to deliver such purpose, but the challenge is to achieve it in such a way not to reduce the unit efficiency. Installation of other plants as gas power or pump-storage hydropower can also partly contribute to the flexibility. This option is especially important in Northern China, where a large portion of thermal plants are CHP with less flexibility in winter heating season. A significant future issue is that as more variable generation comes online dispatchable generation may be displaced thus reducing the available flexibility. Ensuring that future power plants can maintain stable and profitable operation at output levels lower than at present will help to address this issue, but system operators will need to monitor the dynamic stability of the power system.

In parallel with increasing RE penetration, non-renewable clean options, such as nuclear and fossil power with CCS are also in active development in China. Also, synergies and connections also exist between the electricity sector and other energy sectors, for example combining electricity and heat allows for greater flexibility in power systems as thermal storage options are already cost effective. Therefore, for northern China, retrofitting the existing heat system with thermal storage facility can improve the power system flexibility and facilitate the integration of wind power.

- **Demand side measures**

Flexible elements of demand can be used to aid system operation. However, at present the implementations tend to be proprietary in nature and with limited demand controllability actually offered. The development of smart grids and smart appliances, with smart meters linked to control centers, offers the potential to access much greater levels of flexibility from demand. One of the key potential is from household demand. Through real-time pricing, electricity users can be provided incentives to modify or reduce the consumption. Some industry process as ice production, aluminum smelting, water desalination etc. can offer a similar flexibility. EV represents an emerging load, and is potentially a further example of a discretionary load that can be controlled to assist in daily power system operation.

A market or incentive system is required to deliver the demand-side measures. However, unlike in developed countries where demand side management (DSM) has developed into a big market, its development in China is not encouraging at all. An effective tariff scheme that can reflect the resource scarcity and provide correct price signal is a prerequisite.

- **Energy storage**

The amount of energy that can be converted into electricity and then converted back into stored energy, called electricity energy storage, is currently much modest. The most common form of large-scale storage that is mature is the pumped hydro storage. However, pumped storage will result in net energy loss and currently there are no other cost effective options available at utility-scale, though there are vast possibility for novel technologies as hydrogen. When using storage to assist the integration of variable generation, storage should be viewed as a system asset to balance variability. Large-scale development of energy storage at the present time, however, remains questionable due to the generally high capital cost and inherent inefficiency in operation, unless these costs and inefficiencies can be justified through a reduction in curtailment, better use of other flexible resources or more efficient operation of the system.

- **Improved operational/market and planning methods**

In China, existing operational and planning procedures are largely based around dispatchable generation and predictable load patterns. In order to cope with increased penetrations of variable and uncertain generation, however, there is a greater need to identify sources of flexibility in the system operation, to develop probabilistic operations and planning tools and to develop more advanced methods to maintain system stability. Moreover, real-time operations and long-term planning have traditionally been viewed as separate, decoupled activities. With high RE penetrations, the two processes must come closer such that a system is planned that can actually be operated in an economic and reliable manner. In turn, it calls for more accurate forecasting.

Moving to larger balancing areas, or shared balancing between areas, is also desirable. In China, an emergent issue is to develop truly regional power market, instead of the current province-based market.

Advanced planning methods are also required to optimally plan the expansion and upgrade of the power systems to ensure that variable generation can be connected in an efficient manner, especially considering that large geographical and remote areas that will always be involved in China. Detailed modeling of all sources of flexibility will be required, including generation and demand response, such that planning studies reflect the operational potential. Novel planning method, as integrated resource strategic planning (IRSP) can be further developed to include regional analysis function and be used for an appropriate planning tool [17]. Institutional arrangements on the implementing body and procedure need to be changed to effectively address the difficulty of coordination during the planning process. Effective operation and management of large numbers of generation units will be very challenging and require a sophisticated information, communication and control infrastructure. In turn, it calls for radical development of novel smart grids applications and infrastructure.

4. Policy Implications for the Transiton

4.1. Basic Assumptions on Policy Design

In analyzing the failure of the UK Government in promoting sustainable energy, Mitchell (2008) highlighted the importance of the Regulatory State Paradigm and the related policies on innovation, which are vital to a sustainable energy transition [18].

According to Mitchell, the challenge of successfully achieving a transition to a sustainable energy system, rests on the ability of policy makers (at all levels and in all positions) to encourage and enable the necessary changes or innovations at the energy system level, at a firm level, and also in the patterns of sustainable consumption and behaviors across society. However, the logic of market rests on cost-efficiency and may obstruct innovation when the price is uncompetitive.

Government action should be focused on establishing a selection environment which is conductive to 'innovation' and to try to 'channel' innovation as far as possible in the right 'direction'. Particularly because 'not all innovation is good', the Governments cannot simply 'leave' it to market that is blind to technology and fuel, but do have to make a choice about what kind of energy future they want. That is why a market approach alone is not enough and regulations are necessary. Larsson (2009) also argued that market economy is very efficient for the allocation of resources, but in a slower evolution process [1].

Also sometimes the market is not perfect because of "path lock-in". However because of limited time and resources, we need to rapidly expand the use of existing energy-efficient technologies and products and develop a number of new alternatives, which is rightly the advantage of planning. Inslee and Hendricks (2008) argued that the energy transition for the US needs a planned program similar to the American Apollo program that put a man on the Moon in 1969 [19]. Weiss and Bonvillian (2009) also argued for a federal program to stimulate innovation in energy technology in the US [20].

4.2. Policy Design Implications for China

In considering the policy fault-line for innovation, a natural question is how to manage the energy transition process. Therefore, a clear understanding of the innovation process is important. According to the standard pipeline model in innovation theory, a typical innovation process is as follows: research that stems from the curiosity of scientists or inventiveness of engineers leads to an invention, which requires developments to ready it for prototyping and then for commercialization. This model highlights the process of technology supply-push, in which new technologies evolve and push themselves into the marketplace. The major obstacle to innovation of this kind, namely the "valley of death" in the literature, is the gap in support and financing between basic research and later-stage development. As a result, the success of many major innovations has typically depended on a strong injection of public money enabling them to bridge the valley of death [20]. On the reverse side, according to market-pull model, most new products stem from the more mundane process of market or demand pull, during which a market opportunity gives rise to an innovation, which eventually creates requirements for the development and lastly for research. However, the innovators make these long-term investments on new technologies in research, development, and demonstration only when they are convinced that the new environment is here to stay, at least long enough for the resulting technologies to be preferred and come to the market. While these two theories address on the process by which innovation occurs and the external influences to which it responds, the third theory, innovation organization addresses the management of innovation and the organizations in which it takes place. Accordingly, organization mechanisms are needed to help bridge the gaps between public and private sectors and institutions are needed to help smooth the interaction between public, private and academic sectors. However, as is discussed in Section 3.3, energy poses multiple challenges to the models in innovation in that the solution to energy transition is a vast and complex array on both the supply and demand sides while the powerful entrenched array of incumbents is resistant to change.

Regarding energy transition management, Larsson [1] proposed a four-step framework, in which analysis of the overall situation and possible solutions is the first step; in the second step a high-level map of the general direction of change, or strategy is formulated; then in the third step numbers of plans, varying on the stages of the different technologies, are developed based on the strategy; and finally is the managed change. Weiss and Bonvillian [20] also proposed a four-step framework for energy revolution as: firstly, categorize the many and diverse technologies according to their varied launch paths; secondly, match launch paths to policy packages; thirdly, identify the institutional gaps, and finally fill in the gap. The former is largely top-down while the latter is more bottom-up. Trying to combine their merits together we propose an interactive framework for power transition management in China (Figure 9-1).

For the top-down line, first of all a long-term vision for electricity in China will be drafted to guide IRSP power planning. The overall planning then in turn guides more practicable sector plans and detailed implementations, for example, renewable generation planning for the next three decades, Smart Grid planning and its stage by stage implementation, energy efficient building programs, etc. On the bottom-up line, proper policy packages need be designed to support the implementation of sector planning and programs. Whether the policy for a specific energy technology is in the front end support or back end incentive and regulations depends the scientific technology path appraisal based on the state-of-art technology and market development. It should also be noticed that these designed policy packages, will in turn change the resource availability, basic science advance and technology potential on one hand, landscape and regime for energy, industry policy etc. on the other hand, which in turn, will add new insights for updating the long-term vision.

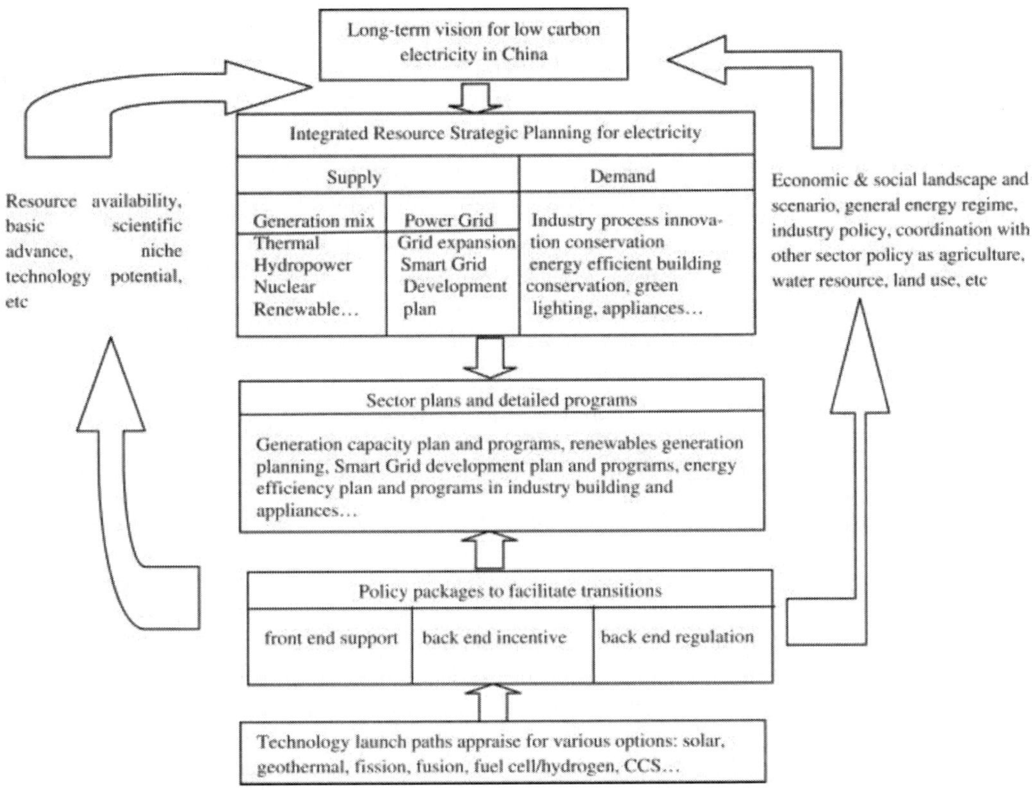

Figure 9.1. An interactive framework for managing energy system transition in China.

According to innovation theory, "front end support", including research and development (R&D), prototyping and demonstration (P&D), public-private partnerships (PPP), monetary prizes for innovators, support for technical education and training are vital for supply-push innovation, while "back end support", economic incentives and regulatory requirements or mandates to encourage innovations are vital for market-pull line. The back end economic incentive includes tax incentives and credits, loan guarantees and low-cost loans, price guarantees, government procurement programs, new-product buy-down programs and general and technology-specific intellectual property policies. On the back end regulation side, policy package includes government standards as appliance standards, energy technology (efficiency) standards in the building and construction sectors, regulatory mandates, such as renewable portfolio standards for utilities, fuel standards, fuel efficiency standards and emission taxes, etc. For all kinds of technologies, the role for the government to implement transitional R&D to overcome the Death Valley at different launch stages is inevitable. Besides, there is substantial role for government in R&D to facilitate the secondary and incremental innovations (or the technology learning) in the manufacturing process.

Though the purpose of the paper is not a case-by-case technology appraisal and policy recommendations as done in previous work, here some technologies will be briefly discussed to demonstrate how to design policy packages according to the state and nature of the specific technology. Five representative technology options are chosen in this section, namely large-scale wind/solar, small-scale off-grid wind/solar distributed generation, energy efficient building, CCS and nuclear technologies.

From many perspectives, China achieved great success in developing wind power from small scale off-grid niches gradually into the mainstream power regime. However, there are many obstacles that prevent the acceleration of wind power and deserve special attention for policy design. For solar power because of the still high cost (PV or CSP) the generators have no incentive to install it. On the other side, without substantial learning-by-doing, the cost will not go down along the learning curve [21]. Because solar is definitely the next most promising renewable technology and because low carbon transition needs an ecosystem rather than one ultimate technology [22], more front end R&D and P&D support for solar is needed for its takeoff in China.

As argued before, to reduce supply cost as well as to develop renewable to the upper most, the optimum paths for renewable generation should include both large-scale on-grid and small-scale off-grid generation. Considering the future reconfiguration scenario of numerous renewable units in different sizes and different technologies located almost everywhere, the Government needs to work out the timetable and program for DG development. However, because of the potential competition with the two national Grid operators, the strongest opposition may come from them. Therefore the Government needs to formulate electricity pricing reform to separate the current jumbled electricity price into generation, transmission, distribution and retail prices. Simultaneously, in some isolated areas with convenient condition (i.e., with plenty of hydropower resources to serve as a backup for renewables and with plenty of solar and wind resources) may be chosen for experimentation and demonstration of DG technology. Besides, DG should best be invested in by a myriad of private (including foreign) investors to reduce the heavy financial burden on the large incumbent. A level investment environment is needed to attract private investors.

Many energy efficient building technologies are mature in a commercialization sense, but their implementation in China is rare. Various sources have confirmed that buildings are a large energy consumer and carbon emitter and the current energy intensity in China's buildings can be cut by a factor of 1/2 to 2/3. It is also worthwhile noticing that because of the low turnover of house, retrofitting of old houses is at least as important as building more efficient new ones. Although improvement on this scale is a dramatic challenge, the conservation opportunities are legion: the current available options including high-efficient bulbs, daylight and actively controlled window shading, highly insulated windows, pigments for roofing and walls with high reflectivity and emissivity to minimize heat gain, heat pump and many others; the newer approaches still need further R&D for passive systems including LEDs, windows with controllable optical and thermal properties, new materials for building shells with superior structural and insulating properties and others. Adoption of these technologies will inevitably result in additional cost to builders and homeowners. Unfortunately, the building sector is notoriously slow to adopt innovations, while both builders and purchasers of real estate favor low initial cost to minimum life-cycle costs and thus insist on short payback periods on investment in conservation that raises initial prices but saves money in the long run. Demonstrations of energy-saving technologies at a scale invested in by government may assist in educating the market, including builders, service providers and consumers. New appliance and lighting standards that fit technology advances in these areas also require implementation. Though energy service in many industrialized countries has developed into an expanding sector, their development in China is still immature. Creating incentives for the Grid Company, especially their urban distribution affiliations to promote energy savings, instead of simply rewarding them for expanded power consumption, could promote their introduction. This will in turn call for the redefinition of the function of the Grid Company, especially the division of the natural monopoly aspects of power transmission & distribution and system operation & dispatch, with the competitive aspects of power supply and energy services for reconfiguration of Power Grid in the future.

CCS is the leading mechanism offering the prospect of allowing continued use of coal while resolving coal's serious CO_2 emission problem. Considering the coal-dominated primary energy and generation structure, CCS is imperative for China. Currently there are about 100 CCS projects in operation globally while only less than 10 of them are implemented at a large scale (one Mt annually). However none of these projects can provide an adequate level of assurance in storing very large quantities of CO_2 over the long term without significant risk of escape, especially given the scale of storage required: a 500-MW coal power plant will generate a billion barrels of liquefied CO_2. China's CCS technology research was initiated by the Ministry of Science and Technology in 2007 and only three small-scale pilot projects were implemented by the end of 2010. The major challenges for CCS are: 1) cutting the cost of CO_2capture, the current capture, transportation and storage process is estimated to increase the power generation cost by 50-100% in China; 2) establishing scientific certainty regarding the security of extremely long-term storage of CO_2in a geological formation at the very large scale required; 3) developing the best practices for operating storage fields long term; 4) developing a (global and national) regulatory structure that would permit the introduction of CCS [23]. For large scale application of CCS in China, more challenges could be added to the list: 5) providing the needed sites of subterranean structures appropriate for long-term carbon storage; 6) building the vast infrastructure for CO_2transportation; and 7) providing the gigantic investment to meet the project requirements (China and India together need 1.17 trillion US$ into 2050 for capture facilities alone [24]. Since CCS projects are pure cost for developers before new ways of commercialized application of captured CO2 can be secured, the generators are unlikely to greet the huge cost imposition of CCS with enthusiasm. Similarly, since developing CCS projects in one country is providing a global public good, without an international legal framework, technology transfer and proper funding support from developed to developing countries like China are unlikely to promote it with enthusiasm. However, the current international arrangements are unable to provide such incentives. Assuming that such mechanisms are effective in the future, to promote CCS, on the front end, perhaps in an international cooperation framework, basic R&D on efficient capture, subterranean structure siting and operation, transportation and perhaps commercialization of CO2 capture could be performed by direct government investment or PPP mechanism. Research is also required to describe, model, predict and monitor the path of injected CO2 in various geological formations and the hydromechanical, chemical and biogeochemical processes involved. The earlier demonstration of CCS technology also needs government investment. At C&P stage, back end mandates such as low carbon energy supply quota, compulsive technology standards for coal plants, a global carbon market and carbon pricing mechanism etc. are needed.

Table 9-4. Policy packages for energy system transition and illustrations

Technology launch path	Research & Development	Prototyping & Demonstration	Investment & Engineering	Commercialization & Popularization	Large scale delivering
	Macro-policies, carbon charge, resource tax, level playing field policy in both technology and investor sense. Consistent policy overview, transitional and technology learning R&D support				
Policy type/ Representative technology analysis	Front end		Back end		
	National basic research activity; technical education and training;	P&D by government fund; PPP; SOE direct investment	PPP; Loan guarantee and low cost loan; Tax credit	Incentive policy Demand side technologies: customer education/ information; purchase subsidy; new-product buy down program Supply side technologies: price guarantee; government procurement; intellectual property policies; grid investment and cost allocation policy for renewable in-access Mandate: government standards as appliance standards, energy technology (efficiency) standards; regulatory mandates as renewable portfolio standard, SOE direct investment, fuel standards, fuel efficiency standards and emission taxes	
Large scale wind/solar project	R&D support: large capacity and efficient wind turbine; wind technology at low wind speed; breakthrough in blade material; breakthrough in CSP and other advanced solar technology P&D support: demonstration of new technology I & E support: policy supporting domestic manufacturing; PPP in technology transfer; low cost loan or tax credit for renewable manufacturers C & P support: integration plan of generation mix and power grid; renewable generation portfolio requirement for generators; grid-access requirement of grid operators; T& D system investment incentive for grid operator and cost allocation policy; favorable dispatch policy for renewable; Feed-in price based on learning curve				
Small scale wind/solar distributed generation projects	P & D support: demonstration for DG C & P support: regulatory mandates and technology standards for DG; clear generation, transmission, distribution and retail pricing mechanism; fair investment environment encouraging small investors				
Energy efficient building	R&D support: transitional R & D on energy efficient building; training of professionals P&D support: demonstration of new energy efficient building technology C & P support: building code; appliance and lighting technology standard; new product price buy down policy; subsidy for house-owner to house retrofitting ; requirement on grid operators to provide energy management services				
CCS	R&D support: basic R & D on CCS technology; transitional R & D to decrease cost P&D support: demonstration of promising CCS technology in power plant; direct SOE incumbent generators investment C & P support: global carbon pricing mechanism; technology and fund support from developed to developing countries; compulsory technology standard; low carbon energy supply quota; tax credit and low cost loan; CDM mechanism				
Nuclear power	R&D support: basic R & D on GIF nuclear and fusion technology P&D support: demonstration of proved new nuclear technology; direct SOE incumbent generators investment; C&P support: favorable investment policy for none-fossil energy; domestic manufacturing policy; standardized design practice ; legal mandate for safety operation				

Back end measures such as government subsidies, PPP, tax credit and CDM mechanisms are also needed to provide the necessary incentives.

Nuclear power has been actively researched and developed in China, although China is a late-comer in this field. The current most advanced nuclear technology, AP1000 was introduced in China in 2007 and the current units under construction are mostly third generation technology, which offers better fuel technology and passive safety so that the reactors shuts down without operator intervention in case of an accident. In 2001 China participated in the fourth generation (GIF) project, which promised to provide the next generation nuclear technology for sustainable energy generation, minimize the nuclear waste, offer life-cycle cost advantage over other energy sources and have more safety and reliability in operations[25]. Six next generation systems, including Sodium-cooled Fast Reactor (SFR), Very High Temperature Reactor (VHTR), Super-Critical Water-cooled Reactor (SCWR), Gas-cooled Fast Reactor (GFR), Lead-cooled Fast Reactor (LFR) and Molten Salt Reactor (MSR) are under R&D by GIF, among which VHTR could be used for hydrogen generation and thus provide a promising position for nuclear in the future hydrogen economy. Because of the serious accident in Japan, a more collaborative international partnership to disseminate best practices on design, operation and disposal of radioactive waste is needed for global nuclear power development. China needs to advance R&D with more active international partnership on the front end, and needs to promote P&D with PPP and tax credit incentives, provide favorable policy for investment in non-fossil energy, promote standardized design practice and domestic manufacturing, and especially important legal mandates for safe operation on the back end (refer to Table 9-4 for a brief summary on discussions in this section).

CONCLUSION

Recognizing the urgency and magnitude of low carbon transition in China, this chapter addresses the topic of energy system transition employing socio-technical transition theory, with an emphasis on China's power sector. By employing the viewpoints of innovation theory, we propose an interactive transition management framework consisting of a top-down line of vision as guidance, integrated planning formulation and sector planning and detailed program implementation, and a bottom-up line as a policy package design based on case-by-case technology launch paths appraised to support sector planning and program implementation, and the central role of policy package designs, because of its feedback effect on macroscopic landscape variables and then into the long-term vision itself.

We argue that our approach both contributes to theoretical and methodological debates on specifying transition pathways, and will be useful in informing policy-makers and other stakeholders.

Regarding China's transition to RE based low carbon energy system, according to the study in the chapter, the following points are extracted to conclude the book:

- A clear energy development strategy and scenario is needed for guiding energy system development in China.

- Integrated planning, covering both generation and grid infrastructure, taking not only the physical power plants but also EPPs from the demand-side into consideration should be drafted for guiding low-level plans and programs.
- Looking into 2020, the most promising pathways from now are transformation including mostly demand-side options, substitution pathways including current mature or would-be mature options such as wind, solar, gas and nuclear power, and reproduction pathways including CHP and UVC to increase efficiency; into 2030 and beyond, more options such as GIF nuclear, numerous DG technologies and fuel cells etc. will add to the substitution list and begin to gradually reconfigure the landscape of power systems into complicated multi-nesting structures (and hence guiding the direction of smart grid development).
- Leaving aside other proposals, our study also reveals institutional gaps (and requirements) for delivering a low carbon transition in China's power sector as: 1) a government body to work out an integrated energy strategy, a sector plan and detailed programs ; 2) a gradual redefinition of the functions of the Grid Company and possible reformation requirement according to transition process and needs; 3) an energy R&D body as well as industry-government consortia to provide integrated translation research and strong front end support and 4) a road mapping think-tank to develop and update the innovation roadmap, combing world-wide industrial, government and academic expertise to assess technologies and to identify obstacles to launch and recommend appropriate policies and a national corporation to carry out financing and demonstration of needed projects.

REFERENCES

[1] Larsson, M., 2009. *Global Energy Transformation*. New York: Palgrave Macmillan press; 2009.

[2] Pacala S.; Socolow R. Stabilization wedges: solving the climate problem for the next 50 years with current technologies. *Science*, 2004 305(13), 968-972.

[3] James H.; Williams et al. The technology path to deep greenhouse gas emissions cuts by 2050: The pivotal pole of electricity. *Science*, 2012 335 (53),53-58.

[4] Yuan, J.; et al. China's 2020 clean energy target: Consistency, pathways and policy implications. *Energy Policy*, 2013. Available from: http://dx.doi.org/10.1016/j.enpol.2013.09.061.

[5] Yuan J; Kang J; Yu C; Hu Z. Energy conservation and emissions reduction in China—Progress and prospective. *Renewable and Sustainable Energy Reviews*, 2011 15(9), 4334-4347.

[6] Yuan J, et al. Managing power system transition in China. Reprinted from *Renewable and Sustainable Energy Review*, 2012 16, 5660-5677. Copyright (2013), with permission from Elsevier.

[7] Chen, Y; Yi J. China Sped up Accessing to "Natural Gas Energy Era". *In Annual Report of China's Energy Development 2010*. Beijing; Social Science Archive Press; 2010.

[8] Rai V.; Victor DG; Thurber MC. Carbon capture and storage at scale: Lessons from the growth of analogous energy technologies. *Energy Policy*, 2010 38, 4089–4098.

[9] IEA (2009). Technology Roadmap-Carbon Capture and Storage, IEA/OECD, Paris.

[10] Smith, A.; Stirling A.; Berkhout, F. The governance of sustainable socio-technical transitions. *Research Policy*, 2005 34(10), 1491-1510.

[11] Rip, A.; Kemp, R. Technological Change. In: S. Rayner and E.L. Malone, editors. *Human Choices and Climate Change vol. 2*. Columbus, Ohio; Battelle Press; 1998.

[12] Shackley, S.; Green, K. A conceptual framework for exploring transitions to decarbonised energy systems in the United Kingdom. *Energy*, 200732(3), 221-236.

[13] Geels, F. W. and Schot J., 2007. Typology of sociotechnical transition pathways. *Research Policy*, 200736 (3), 399-417.

[14] Zhao, X.; Wang, F.; Wang M. Large-scale utilization of wind power in China: Obstacles of conflict between market and planning. *Energy Policy*, 2012 48, 222–232.

[15] SimsR; et al. Integration of Renewable Energy into Present and Future Energy Systems. *In IPCC Special Report on Renewable Energy Sources and Climate Change Mitigation.* Cambridge, United Kingdom and New York, NY, USA; Cambridge University Press; 2011.

[16] SGCC (2009). Smart Grids Development of SGCC during 12th Five-Year-Plan period.

[17] Hu, Z. et al. Integrated resource strategic planning in China. *Energy Policy*, 201038(8), 4635-4642.

[18] Mitchell, C. *The Political Economy of Sustainable Energy*. New York; Palgrave Macmillan press; 2011.

[19] Inslee, J.; Hendricks B. *Apollo's Fire*. Washington; Island Press, 2008.

[20] Weiss, C. and Bonvillian, W. B. *Structuring an Energy Technology Revolution*. Cambridge, Massachusetts; The MIT Press, 2009.

[21] Zwaan, B. and Rabl A. The learning potential of photovoltaics: implications for energy policy. *Energy Policy*, 200432 (13), 1545-1554.

[22] Friedman, Thomas L. Hot, Flat and Crowded: Why We Need a Green Revolution and How Can It Renew America. New York; Farrar, Straus and Girous; 2008.

[23] MIT. The Future of Coal.2006. Available from: http://web.mit.edu/coal/.

[24] Almendra, F., et al. CCS Demonstration in Developing Countries: Priorities for a Financing Mechanism for Carbon Dioxide Capture and Storage, WRI Working Paper. World Resources Institute, Washington DC.2011. Available from: http://www.wri.org/publication/ccs-demonstration-in-developing-countries.

[25] OECD/IEA and OECD/NEA. Nuclear energy technology roadmap.2010.Available from: http://www.iea.org/papers/2010/nuclear_roadmap.pdf.

EDITORS' CONTACT INFORMATION

Dr. Jiahai Yuan
Associate Professor, Ph.D.
School of Economics and Management
North China Electric Power University
Shuangbei Rd, Daxing, Beijing, China
yuanjh126@126.com

Dr. Guoliang Luo
Associate Professor, Ph.D.
School of Economics and Management
North China Electric Power University
Shuangbei Rd, Daxing, Beijing, China

Dr. Jianguo Chen, Ph.D.
School of Humanities and Social Science
North China Electric Power University
Shuangbei Rd, Daxing, Beijing, China

INDEX

#

A

B

C

D

E

F

G

H

I

N

O

P

Q

R

S

Y